Self-Similarity and Multiwavelets in Higher Dimensions

[redacted]

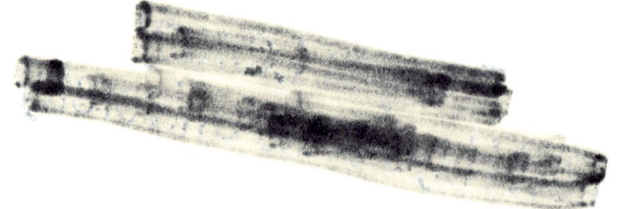

MEMOIRS of the American Mathematical Society

Number 807

Self-Similarity and Multiwavelets in Higher Dimensions

Carlos A. Cabrelli
Christopher Heil
Ursula M. Molter

July 2004 • Volume 170 • Number 807 (end of volume) • ISSN 0065-9266

American Mathematical Society
Providence, Rhode Island

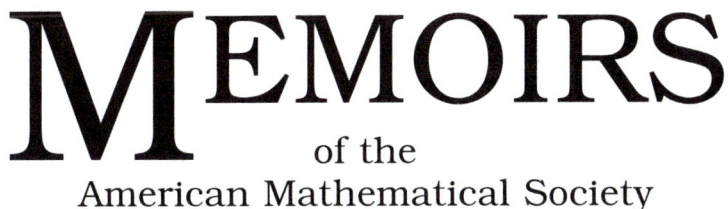

2000 *Mathematics Subject Classification.*
Primary 39A10; Secondary 39B62.

Library of Congress Cataloging-in-Publication Data

Cabrelli, Carlos A, 1949–
Self-similarity and multiwavelets in higher dimension / Carlos A. Cabrelli, Christopher Heil, Ursula M. Molter.
p. cm. — (Memoirs of the American Mathematical Society, ISSN 0065-9266 ; no. 807)
"Volume 170, number 807 (end of volume)."
Includes bibliographical references.
ISBN 0-8218-3520-3 (alk. paper)
1. Difference equations. 2. Functional equations. 3. Inequalities (Mathematics) I. Heil, Christopher, 1960– II. Molter, Ursula M., 1957– III. Title. IV. Series.
QA3.A57 no. 807
[QA431]
510 s–dc22
[515′.625] 2004046102

Memoirs of the American Mathematical Society

This journal is devoted entirely to research in pure and applied mathematics.

Subscription information. The 2004 subscription begins with volume 167 and consists of six mailings, each containing one or more numbers. Subscription prices for 2004 are $583 list, $466 institutional member. A late charge of 10% of the subscription price will be imposed on orders received from nonmembers after January 1 of the subscription year. Subscribers outside the United States and India must pay a postage surcharge of $31; subscribers in India must pay a postage surcharge of $43. Expedited delivery to destinations in North America $35; elsewhere $130. Each number may be ordered separately; *please specify number* when ordering an individual number. For prices and titles of recently released numbers, see the New Publications sections of the *Notices of the American Mathematical Society.*

Back number information. For back issues see the *AMS Catalog of Publications.*

Subscriptions and orders should be addressed to the American Mathematical Society, P. O. Box 845904, Boston, MA 02284-5904, USA. *All orders must be accompanied by payment.* Other correspondence should be addressed to 201 Charles Street, Providence, RI 02904-2294, USA.

Memoirs of the American Mathematical Society is published bimonthly (each volume consisting usually of more than one number) by the American Mathematical Society at 201 Charles Street, Providence, RI 02904-2294, USA. Periodicals postage paid at Providence, RI. Postmaster: Send address changes to Memoirs, American Mathematical Society, 201 Charles Street, Providence, RI 02904-2294, USA.

This publication is indexed in *Science Citation Index®*, *SciSearch®*, *Research Alert®*, *CompuMath Citation Index®*, *Current Contents®/Physical, Chemical & Earth Sciences.*
Printed in the United States of America.

∞ The paper used in this book is acid-free and falls within the guidelines established to ensure permanence and durability.
Visit the AMS home page at `http://www.ams.org/`

10 9 8 7 6 5 4 3 2 1 09 08 07 06 05 04

Contents

Abstract

Let A be a dilation matrix, an $n \times n$ expansive matrix that maps a full-rank lattice $\Gamma \subset \mathbf{R}^n$ into itself. Let Λ be a finite subset of Γ, and for $k \in \Lambda$ let c_k be $r \times r$ complex matrices. The refinement equation corresponding to A, Γ, Λ, and $c = \{c_k\}_{k \in \Lambda}$ is $f(x) = \sum_{k \in \Lambda} c_k \, f(Ax - k)$. A solution $f \colon \mathbf{R}^n \to \mathbf{C}^r$, if one exists, is called a refinable vector function or a vector scaling function of multiplicity r. In this manuscript we characterize the existence of compactly supported L^p or continuous solutions of the refinement equation, in terms of the p-norm joint spectral radius of a finite set of finite matrices determined by the coefficients c_k. We obtain sufficient conditions for the L^p convergence ($1 \leq p \leq \infty$) of the Cascade Algorithm $f^{(i+1)}(x) = \sum_{k \in \Lambda} c_k \, f^{(i)}(Ax - k)$, and necessary conditions for the uniform convergence of the Cascade Algorithm to a continuous solution. We also characterize those compactly supported vector scaling functions which give rise to a multiresolution analysis for $L^2(\mathbf{R}^n)$ of multiplicity r, and provide conditions under which there exist corresponding multiwavelets whose dilations and translations form an orthonormal basis for $L^2(\mathbf{R}^n)$.

Received by the editor May 21, 1999; revised June 30, 2003.

1991 *Mathematics Subject Classification.* Primary 39A10; Secondary 39B62.

Key words and phrases. Cascade Algorithm, Dilation matrix, joint spectral radius, multiresolution analysis, multiwavelets, nonseparable wavelets, refinement equations, refinable functions, scaling functions, self-similarity, wavelets.

Acknowledgments

It is a pleasure to thank some of the people who have assisted us with this manuscript. In particular, our thanks go to Ana Ruedin for computing the coefficients for the scaling functions discussed in Chapter 5, to Yang Wang for many conversations about tilings and related topics, and to Denise Jacobs and Maria Luisa Gordillo for careful readings of the entire manuscript. Finally, we express our deep appreciation to Luis Caffarelli, Carlos Kenig, and Palle Jorgensen for valuable editorial advice and assistance.

A portion of the research for this paper was performed during visits by Cabrelli and Molter to the School of Mathematics at the Georgia Institute of Technology. These authors thank the School for its hospitality and support during these visits. the work of Heil was partially supported by National Science Foundation Grants DMS-9401340, DMS-9970524, and DMS-0139261. The work of Cabrelli and Molter was partially supported by Grants UBACyT EX48 and TW84, CONICET PIP456, and ANPCyT PICT 03134/98

CHAPTER 1

Introduction

1.1. Description of Results

Let $\Gamma \subset \mathbf{R}^n$ be a full-rank lattice (the image of $\mathbf{Z}^d$ under an invertible matrix). Let A be a dilation matrix, i.e., A is an expansive $n \times n$ matrix which maps Γ into itself. Let Λ be a finite subset of Γ. Then given $r \times r$ matrices c_k for $k \in \Lambda$, the refinement equation associated to A, Γ, Λ, and $c = \{c_k\}_{k\in\Lambda}$ is

$$f(x) \;=\; \sum_{k\in\Lambda} c_k\, f(Ax-k), \qquad x \in \mathbf{R}^n, \tag{1.1}$$

where a solution f, if one exists, is a vector-valued function $f\colon \mathbf{R}^n \to \mathbf{C}^r$, i.e.,

$$f(x) \;=\; \begin{bmatrix} f_1(x) \\ \vdots \\ f_r(x) \end{bmatrix}.$$

We call a compactly supported solution of the refinement equation a *refinable (vector) function* or a *(vector) scaling function*, and r is its multiplicity.

In this manuscript we will characterize the existence of compactly supported L^p or continuous solutions of the refinement equation. The *Cascade Algorithm* is the iteration

$$f^{(i+1)}(x) \;=\; \sum_{k\in\Lambda} c_k\, f^{(i)}(Ax-k). \tag{1.2}$$

We obtain sufficient conditions for the L^p convergence ($1 \le p \le \infty$) of the Cascade Algorithm, and necessary conditions for the uniform convergence of the Cascade Algorithm to a continuous solution. We also characterize when a solution of a refinement equation is a generator of a multiresolution analysis (see Definition 4.1) for $L^2(\mathbf{R}^n)$. Such a generator enables the construction of multiwavelet orthonormal bases for $L^2(\mathbf{R}^n)$.

The higher-dimensional setting of this manuscript, allowing an arbitrary dilation matrix, creates significant geometrical obstacles to the analysis of the refinement equation. In Chapter 2 we prove a number of technical lemmas and develop a set of geometrical tools which are needed to prove the main results of Chapters 3 and 4. In particular, we prove that the support of the scaling function is a compact set that is contained in the attractor of an iterated function system (IFS) determined by the set Λ (Theorem 2.2).

Given a choice of dilation matrix A and a choice of digits D (a set of representatives of $\Gamma/A\Gamma$), there exists a unique compact set Q that is the attractor of another IFS determined by A and D. Except for certain dilation matrices in dimensions 4 and higher, there exists a choice of digits for which this attractor Q tiles $\mathbf{R}^n$ with overlaps of measure zero using translations by Γ (see Theorem 2.3). We assume this is the case for the dilation matrices considered in this manuscript. Although the tile Q typically has a fractal boundary, we prove in Proposition 2.10 that there exists a subset $\tilde{Q}$ of Q that tiles $\mathbf{R}^n$ using translations by Γ without overlaps.

We transform the refinement equation to an equivalent vector equation over the tile in Proposition 2.13 and Corollary 2.15. This will allow us in Chapter 3 to analyze the convergence of the Cascade Algorithm in terms of the spectral properties of a finite set of matrices. To this end, in Proposition 2.17 we derive lower and upper bounds for the p-norm joint spectral radius of a set of matrices in terms of an appropriate matrix norm.

In Theorem 3.1 we prove the existence of a fixed point of a general class of functional equations. The solutions of these equations are called generalized self-similar functions. The refinement equation is a particular member of this class.

In Theorem 3.4 we give sufficient time-domain conditions for the existence of a unique continuous or L^p vector scaling function in terms of the p-norm joint spectral radius ($1 \le p \le \infty$) of a finite set of finite matrices T_d restricted to a specific subspace E_0, all determined by the coefficients c_k. Furthermore, we show that if these conditions are satisfied, then the Cascade Algorithm converges geometrically in L^p to this unique solution.

A vector function $g\colon \mathbf{R}^n \to \mathbf{C}^r$ has *accuracy* κ if every polynomial q on $\mathbf{R}^n$ with complex coefficients and $\deg(q) < \kappa$ can be written

$$q(x) = \sum_{k\in\Gamma} a_k\, g(x+k) \quad \text{a.e.}$$

for some $1 \times r$ row vectors a_k. We prove in Theorem 3.17 that if a scaling function f has accuracy κ then the matrices T_d can be simultaneously brought into a particular block triangular form. The subspace E_0 mentioned before is one of the invariant subspaces corresponding to this simultaneous triangularization. This is a key ingredient for obtaining necessary conditions for the existence of a continuous solution to the refinement equation.

In Theorem 3.22 we prove that if a continuous solution to the refinement equation exists which has L^∞-stable translates (see Definition 3.18), then the Cascade Algorithm converges uniformly for the starting function $\chi_{\tilde{Q}}$, where $\tilde{Q}$ is the subset of Q that tiles $\mathbf{R}^n$ without overlaps. In Theorem 3.26 we prove that if the Cascade Algorithm converges pointwise everywhere for the starting function $\chi_{\tilde{Q}}$ to a continuous solution of the refinement equation, then the ∞-norm joint spectral radius of the matrices T_d restricted to E_0 is strictly less than 1. We bound the Hölder exponent of continuity of a continuous scaling function in Proposition 3.27.

In Theorem 4.4 we characterize all compactly supported vector-valued functions with orthonormal lattice translates which generate a multiresolution analysis (Definition 4.1) of $L^2(\mathbf{R}^n)$. In particular, any such function is a solution of a refinement equation. Once a multiresolution analysis is given, Theorem 4.11 provides conditions under which there exist corresponding *multiwavelets* whose dilates and translates form an orthonormal basis for $L^2(\mathbf{R}^n)$.

Finally, in Chapter 5, we apply the results of this manuscript by numerically constructing new examples of continuous, compactly supported vector scaling functions with orthonormal lattice translates and accuracy $\kappa = 2$ that are refinable with respect to the quincunx dilation $A = \left[\begin{smallmatrix} 1 & 1 \\ 1 & -1 \end{smallmatrix}\right]$. We also construct the corresponding multiwavelets.

1.2. A Historical Overview

The history of the study of refinement equations is complex, involving researchers from numerous fields and disciplines. We briefly outlinc some of the highlights of that history here, emphasizing those results most directly related to this manuscript. We will not attempt to give an exhaustive summary of all literature related to refinement equations. Additional related papers can be found in the references of the articles that we cite.

Micchelli and Prautzsch [**MP89**] and Daubechies and Lagarias [**DL92**] each independently introduced a time-domain method for testing the smoothness of refinable functions in the one-dimensional, single function case ($n = 1$, $r = 1$). The conditions developed in [**MP89**], [**DL92**] were based on the computation of all possible products of a set of finite matrices directly determined by the coefficients c_k. In particular, Daubechies and Lagarias [**DL92**] rediscovered the uniform joint spectral radius (JSR) of Rota and Strang [**RS60**], and used it as a fundamental tool for formulating these conditions. Many papers, utilizing a variety of techniques, have since studied additional properties of the scaling function, such as Sobolev or Besov space membership, e.g., [**Eir92**], [**Vil94a**]. Of particular relevance to this manuscript are the papers of Y. Wang [**Wan96**], who introduced a 1-norm generalization of the JSR in order to formulate a test for the existence of L^1-solutions to the refinement equation, and Jia [**Jia95**], who independently introduced a p-norm generalization of the JSR to test for L^p-solutions. The p-JSR was also used implicitly by Lau and J. Wang in [**LauW95**].

The above-mentioned papers are all concerned with one-dimensional, single-function refinement equations. Cohen and Daubechies [**CD93**] generalized some of the one-dimensional tests of [**DL92**] to the case of two-dimensional, single function refinement equations using a quincunx dilation matrix. Some results giving tests for the existence of continuous solutions or the Sobolev and Hölder regularity of the solution in the multidimensional, single-function case ($n > 1$, $r = 1$) appear in [**Vil94b**], [**CGV99**], [**Jia99**].

The accuracy conditions for one-dimensional, multi-function refinement equations are considerably more involved than in the single-function case. These conditions were derived independently by Heil, Strang, and Strela [**HSS96**] and by Plonka [**Plo97**]. Plonka further discovered that these accuracy conditions imply a factorization of the matrix-valued symbol of the refinement equation (the Fourier transform of the sequence of matrix coefficients $\{c_k\}$). This factorization is not as convenient as in the single-function case, but it has been been useful for the construction and analysis of multiwavelets in one dimension [**MS97**], [**CDP97**]. The accuracy conditions for the multidimensional, multi-function case were derived in [**CHM98**], [**CHM00**], with some similar results for the case of diagonalizable dilation matrices in [**Jng99**]. The *order of approximation* of f is closely related to its accuracy, but can be distinct in higher dimensions. We refer to [**BDR94a**], [**BDR94b**] and related works for discussions of order of approximation.

There have been a few specific constructions of non-tensor product orthonormal wavelet bases in higher dimensions. Gröchenig and Madych [**GM92**] studied the particular case of higher-dimensional, single-function dilation equations whose solution is the characteristic function of a tile. These special refinement equations yield discontinuous wavelets that are higher-dimensional analogues of the Haar basis for $L^2(\mathbf{R})$. Kovačević and Vetterli [**KoV92**] constructed a single specific example of continuous scaling function on $\mathbf{R}^2$ that is refinable with respect to the quincunx dilation matrix $A = \left[\begin{smallmatrix} 1 & 1 \\ 1 & -1 \end{smallmatrix}\right]$ and whose $\mathbf{Z}^2$-translates are orthonormal (see [**Vil94b**] for the proof that this scaling function is continuous, which we also verify in Section 5.2). This was for many years the only known example of a continuous, nonseparable, two-dimensional, compactly supported orthonormal scaling function. More recent constructions by Kovačević and Vetterli are in [**KoV95**]. Recently, He and Lai constructed some examples and then families of two-dimensional, nonseparable, continuous, compactly supported scaling functions with orthonormal translates that are refinable with respect to the uniform dilation $A = 2I$ [**HL97**]. By choosing a specific geometry for the support Λ of the coefficients c_k, Belogay and Wang [**BW99**] were able to impose a limited factorization of the symbol and use that to construct a specific family of two-dimensional, compactly supported scaling functions with orthonormal translates and increasing regularity that are refinable with respect to the dilation $A = \left[\begin{smallmatrix} 0 & 2 \\ 1 & 0 \end{smallmatrix}\right]$. One-dimensional orthonormal multiwavelets were constructed in [**Alp93**], [**GLT93**], [**GL94**], [**GHM94**], [**DGHM96**]. Donovan, Geronimo, and Hardin constructed two-dimensional multiwavelets that are refinable with respect to the uniform dilation $A = 3I$ [**DGH95**]. Ayache has some constructions using the uniform dilation $A = 2I$ [**Aya99a**], [**Aya99a**]. We also remark on some related constructions with somewhat different properties. Examples of orthonormal, multidimensional wavelets whose Fourier transforms are compactly supported are presented in [**DLS97**], [**Cal99**], [**BL01**]. Compactly supported, multidimensional, biorthogonal wavelets are constructed in [**DM97**], [**Der99**], [**HL99**], [**JRS99**], [**KaV99**], [**KS00**]. Compactly supported, multidimensional wavelet frames are presented in [**Han97**], [**GR98**]. The literature on these topics is of course always expanding; the references given above are typical but not exhaustive.

CHAPTER 2

Matrices, Tiles and the Joint Spectral Radius

2.1. Miscellaneous Notation

We use the conventions $1/\infty = 0$, $1/0 = \infty$, and $0^0 = 1$.

The absolute value of a real or complex number z is denoted by $|z|$. The complex conjugate of z is $\bar{z}$.

The transpose of a matrix B is B^{T}. The Hermitian, or conjugate transpose, is B^*.

The cardinality of a finite set F is denoted by $\#F$.

The interior of a set $E \subset \mathbf{R}^n$ is E°, the boundary of E is ∂E, and the closure of E is $\overline{E}$. If E is measurable, its Lebesgue measure is denoted by $|E|$. The characteristic function of E is denoted χ_E. The Kronecker delta is denoted $\delta_{i,j}$.

The open ball in $\mathbf{R}^n$ of radius $\varepsilon > 0$ centered at $x \in \mathbf{R}^n$ is

$$B(x, \varepsilon) = \{y \in \mathbf{R}^n : \|x - y\| < \varepsilon\},$$

measured with respect to whatever norm on $\mathbf{R}^n$ is currently in force. Most computations in this manuscript are independent of the choice of norm on $\mathbf{R}^n$; if not specifically stated then the norm is taken to be the Euclidean norm on $\mathbf{R}^n$.

The support of a vector-valued function $g = (g_1, \dots, g_r)^{\mathrm{T}} \colon \mathbf{R}^n \to \mathbf{C}^r$ is the closure of $\{x \in \mathbf{R}^n : g(x) \neq 0\}$. Integrals of g are computed componentwise. In particular, if g is integrable then we define its Fourier transform by

$$\begin{aligned}\hat{g}(\omega) &= \int_{\mathbf{R}^n} g(x)\, e^{-2\pi i x \cdot \omega}\, dx \\ &= \left(\int_{\mathbf{R}^n} g_1(x)\, e^{-2\pi i x \cdot \omega}\, dx, \dots, \int_{\mathbf{R}^n} g_r(x)\, e^{-2\pi i x \cdot \omega}\, dx \right)^{\mathrm{T}}.\end{aligned}$$

The space $L^p(\mathbf{R}^n)$ consists of all complex-valued functions f on $\mathbf{R}^n$ for which the norm

$$\|f\|_p = \left(\int_{\mathbf{R}^n} |f(x)|^p\, dx \right)^{1/p}, \qquad \text{if } 1 \le p < \infty,$$

or

$$\|f\|_\infty = \operatorname*{ess\,sup}_{x \in \mathbf{R}^n} |f(x)|, \qquad \text{if } p = \infty,$$

is finite. We use the standard inner product on $L^2(\mathbf{R}^n)$:

$$\langle g, h \rangle = \int_{\mathbf{R}^n} g(x)\, \overline{h(x)}\, dx, \qquad g, h \in L^2(\mathbf{R}^n).$$

Let X be a closed subset of $\mathbf{R}^n$, and let $\|\cdot\|$ be any fixed norm on $\mathbf{C}^r$. Then we define $L^p(X, \mathbf{C}^r)$ to be the Banach space of all mappings $g \colon X \to \mathbf{C}^r$ such that

$$\|g\|_{L^p}^p = \int_X \|g(x)\|^p\, dx < \infty,$$

with the usual modification if $p = \infty$. For simplicity, we define $L^p(X) = L^p(X, \mathbf{C})$. This definition of $L^p(X, \mathbf{C}^r)$ is independent of the choice of norm $\|\cdot\|$ on $\mathbf{C}^r$ in the sense that each such choice yields an equivalent norm for $L^p(X, \mathbf{C}^r)$. If E is a nonempty closed subset of $\mathbf{C}^r$, then $L^p(X, E)$ is the closed subset of $L^p(X, \mathbf{C}^r)$ consisting of functions which take values in E.

We will assume throughout this manuscript that A is a fixed dilation matrix with associated full-rank lattice $\Gamma \subset \mathbf{R}^n$. That is, $A(\Gamma) \subset \Gamma$ and every eigenvalue λ of A satisfies $|\lambda| > 1$. We will consider refinement equations of multiplicity r given as in (1.1), i.e.,

$$f(x) = \sum_{k \in \Lambda} c_k \, f(Ax - k), \qquad x \in \mathbf{R}^n,$$

where Λ is a fixed finite subset of Γ and the c_k are fixed $r \times r$ matrices. A solution of the refinement equation is called a *vector scaling function* or a *refinable vector function.*

The *refinement operator* associated with this refinement equation is the mapping S, acting on vector functions $g\colon \mathbf{R}^n \to \mathbf{C}^r$, defined by

$$Sg(x) = \sum_{k \in \Lambda} c_k \, g(Ax - k), \qquad x \in \mathbf{R}^n. \tag{2.1}$$

A scaling function is thus a fixed point of S. The *cascade algorithm* defined in (1.2) is the iteration

$$f^{(i+1)} = Sf^{(i)}.$$

We will use a generalized matrix notation which allows matrices or vectors to be indexed by arbitrary countable sets. If desired, such generalized matrices can always be realized as ordinary matrices by choosing a specific ordering for the index set. The actual ordering used is not important, as long as the same ordering is used consistently. To be precise, let J and K be finite or countable index sets. Let $m_{j,k}$ be $r \times s$ matrices for $j \in J$ and $k \in K$. Then we say that $M = [m_{j,k}]_{j \in J, k \in K} \in (\mathbf{C}^{r \times s})^{J \times K}$ is a $J \times K$ matrix (with $r \times s$ block entries). If $N = [n_{k,\ell}]_{k \in K, \ell \in L} \in (\mathbf{C}^{s \times t})^{K \times L}$, then the product of the $J \times K$ matrix M with the $K \times L$ matrix N is the $J \times L$ matrix formally defined by

$$MN = \left[\sum_{k \in K} m_{j,k} \, n_{k,\ell} \right]_{j \in J, \ell \in L}.$$

Most summations encountered in this manuscript will contain only finitely many nonzero terms. A "column vector" is a $J \times 1$ matrix, which we will denote by $v = [v_j]_{j \in J}$. The entries v_j may be scalars or $r \times s$ blocks. In particular,

$$\mathbf{C}^r = \mathbf{C}^{r \times 1} = \left\{ \begin{bmatrix} u_1 \\ \vdots \\ u_r \end{bmatrix} : u_1, \dots, u_r \in \mathbf{C} \right\}$$

is the space of column vectors of length r. Analogously, a "row vector" is a $1 \times J$ matrix, which we will denote by $u = (u_j)_{j \in J}$. In particular, $\mathbf{C}^{1 \times r}$ is the space of all row vectors of length r, i.e.,

$$\mathbf{C}^{1 \times r} = \{u^{\mathrm{T}} : u \in \mathbf{C}^r\} = \{(u_1, \dots, u_r) : u_1, \dots, u_r \in \mathbf{C}\}.$$

2.2. Attractors and Tiles

Since $A(\Gamma) \subset \Gamma$, the dilation matrix A necessarily has integer determinant. We define

$$m = |\det(A)|,$$

and let

$$D = \{d_1, \dots, d_m\}$$

be a *full set of digits* with respect to A and Γ, i.e., a complete set of representatives of the order-m group $\Gamma/A(\Gamma)$. Because D is a full set of digits, the lattice Γ is partitioned into the disjoint cosets

$$\Gamma_d = A(\Gamma) - d = \{Ak - d : k \in \Gamma\}, \qquad d \in D. \tag{2.2}$$

Let $\gamma_1, \dots, \gamma_n$ be a set of generators for the lattice Γ, i.e., independent vectors such that

$$\Gamma = \{m_1\gamma_1 + \cdots + m_n\gamma_n : m_i \in \mathbf{Z}\}.$$

Then the rectangular parallelepiped

$$P = \{x_1\gamma_1 + \cdots + x_n\gamma_n : 0 \le x_i < 1\} \tag{2.3}$$

is a *fundamental domain* for the group $\mathbf{R}^n/\Gamma$, and $\mathbf{R}^n$ is partitioned into the sets $\{P + k\}_{k\in\Gamma}$. For example, if $\Gamma = \mathbf{Z}^n$, then we can choose $\gamma_1, \dots, \gamma_n$ so that $P = [0,1)^n$.

2.2.1. Attractors.

The space $\mathcal{H}(\mathbf{R}^n)$ consisting of all nonempty, compact subsets of $\mathbf{R}^n$ is a complete metric space under the Hausdorff metric $h(\cdot,\cdot)$ defined by

$$h(B, C) = \inf\{\varepsilon > 0 : B \subset C_\varepsilon \text{ and } C \subset B_\varepsilon\},$$

where

$$B_\varepsilon = \{x \in \mathbf{R}^n : \operatorname{dist}(x, B) < \varepsilon\}. \tag{2.4}$$

Thus

$$h(B, C) < \varepsilon \quad \Longleftrightarrow \quad B \subset C_\varepsilon \text{ and } C \subset B_\varepsilon.$$

Since all norms on $\mathbf{R}^n$ are equivalent, the definition of the Hausdorff metric is independent of the choice of norm used to measure distance in (2.4).

For each $k \in \Gamma$, let $w_k \colon \mathbf{R}^n \to \mathbf{R}^n$ be the affine map

$$w_k(x) = A^{-1}(x + k). \tag{2.5}$$

Since A^{-1} is contractive, each w_k is a contractive mapping on $\mathbf{R}^n$. For each finite subset $H \subset \Gamma$, define $w_H \colon \mathcal{H}(\mathbf{R}^n) \to \mathcal{H}(\mathbf{R}^n)$ by

$$w_H(B) = \bigcup_{k\in H} w_k(B) = A^{-1}(B + H). \tag{2.6}$$

Using the fact that each w_k is contractive on $\mathbf{R}^n$ under the Euclidean norm, it can be shown that w_H is contractive on $\mathcal{H}(\mathbf{R}^n)$ under the Hausdorff metric [**Hut81**]. The Contraction Mapping Theorem therefore implies that there exists a unique nonempty compact set $K_H \subset \mathbf{R}^n$ such that

$$w_H(K_H) = K_H.$$

That is, K_H is defined by the property

$$K_H = A^{-1}(K_H + H). \tag{2.7}$$

The set K_H is called the *attractor* of the iterated function system (IFS) generated by $\{w_k\}_{k\in H}$ [**Hut81**]. In particular, the attractors K_Λ and $Q = K_D$ of the IFS's generated by $\{w_k\}_{k\in\Lambda}$ and $\{w_k\}_{k\in D}$, respectively, will play important roles throughout this manuscript. Because w_H is a contraction on $\mathcal{H}(\mathbf{R}^n)$, the iteration $K^{(i+1)} = w_H(K^{(i)})$ converges in the Hausdorff metric to the attractor K_H for any nonempty compact starting set $K^{(0)}$. Therefore, any attractor K_H can always be approximated as closely as desired.

We can use (2.7) to obtain another expression for K_H. Iterating (2.7) k times, we see that

$$K_H = \sum_{j=1}^{k} A^{-j}(H) + A^{-k}(K_H).$$

Then, using the fact that A^{-1} is a contraction, it follows that

$$K_H = \sum_{j=1}^{\infty} A^{-j}(H) = \left\{\sum_{j=1}^{\infty} A^{-j}h_j : h_j \in H\right\}. \tag{2.8}$$

The following properties of an attractor K_H will be useful. Parts (a), (b), and (c) of the following lemma are also valid for more general iterated function systems [**Ban91**], while parts (d), (e), and (f) make use of the fact that the functions w_k defined in (2.5) are affine mappings.

LEMMA 2.1. *Let $B \in \mathcal{H}(\mathbf{R}^n)$, and let H, H_1, H_2 be finite subsets of Γ.*

(a) *If $B \subset w_H(B)$, then $B \subset K_H$.*

(b) *If $w_H(B) \subset B$, then $K_H \subset B$.*

(c) *If $H_1 \subset H_2$, then $K_{H_1} \subset K_{H_2}$.*

(d) *$w_H(K_H^\circ) \subset K_H^\circ$.*

(e) *$|\partial K_H| = 0$.*

(f) *If $\gamma \in \Gamma$, then $K_{H+\gamma} = K_H + (A-I)^{-1}\gamma$.*

Next, we prove that a scaling function must be supported in K_Λ.

PROPOSITION 2.2.

(a) *If $g\colon \mathbf{R}^n \to \mathbf{C}^r$ is compactly supported, then* $\operatorname{supp}(Sg) \subset w_\Lambda(\operatorname{supp}(g))$.

(b) *If $f\colon \mathbf{R}^n \to \mathbf{C}^r$ is a compactly supported solution of the refinement equation, then* $\operatorname{supp}(f) \subset K_\Lambda$.

PROOF. (a) It follows from (2.1) that

$$\operatorname{supp}(Sg) \subset A^{-1}(\operatorname{supp}(g) + \Lambda) = w_\Lambda(\operatorname{supp}(g)).$$

(b) If $Sf = f$ then part (a) implies $\operatorname{supp}(f) \subset w_\Lambda(\operatorname{supp}(f))$, so $\operatorname{supp}(f) \subset K_\Lambda$ by Lemma 2.1(a). □

2.2.2. The Tile Q. Since $D = \{d_1, \ldots, d_m\}$ is a full set of digits with respect to A and Γ, if we take any $\gamma \in \Gamma$ then $D+\gamma$ will also be a full set of digits with respect to A and Γ. Further, by Lemma 2.1(f), we have $K_{D+\gamma} = K_D + (A-I)^{-1}\gamma$. Hence we can always translate the digit set D as we like, at the cost of correspondingly translating the set $Q = K_D$, which is the attractor of the IFS generated by $\{w_d\}_{d\in D}$. Without loss of generality, we therefore will always assume that $0 \in D$. Equation (2.8) then implies that $0 \in Q$.

The following properties of Q will be useful [**Ban91**], cf. also [**GM92**].

LEMMA 2.3. *Let $Q = K_D$, and let P be the fundamental domain defined in* (2.3). *Then the following statements hold.*

(a) $Q + \Gamma = \mathbf{R}^n$.

(b) *Q has nonempty interior, Q is the closure of Q°, and $|\partial Q| = 0$.*

(c) *$|Q \cap (Q + k)| = 0$ for all $k \in \Gamma \setminus \{0\}$ if and only if $|Q| = |P|$. In this case, $Q \cap (Q+k) \subset \partial Q$ for each $k \in \Gamma \setminus \{0\}$.*

(d) $\#(Q^\circ \cap \Gamma) \le 1$.

In other words, part (c) above says that if $|Q| = |P|$, then Q is a *tile* in the sense that the Γ-translates $\{Q + k\}_{k\in\Gamma}$ cover $\mathbf{R}^n$ with overlaps of measure zero. A longstanding open problem was the question of whether for each dilation matrix A there exists a full set of digits D such that the corresponding attractor Q is a tile. Lagarias and Wang proved that this is the case if $n = 1, 2, 3$ or if $m = |\det(A)| > n$ [**LagW95a**], [**LagW96**], [**LagW97**]. Potiopa [**Pot97**] recently showed that if $n = 4$ and

$$A = \begin{bmatrix} 0 & 1 & 0 & 0 \\ 0 & 0 & 1 & 0 \\ 0 & 0 & -1 & 2 \\ -1 & 0 & -1 & 1 \end{bmatrix},$$

then there is no set of digits D such that $Q = K_D$ is a tile, cf. [**LagW99**]. Note that this matrix A has determinant 2.

We will only deal in this manuscript with the case where a tile Q exists. Precisely, the following standing assumption will always be in force.

STANDING ASSUMPTION 2.4. We will assume throughout this manuscript that whenever a dilation matrix A and choice of digits D are given, the corresponding attractor $Q = K_D$ is a tile. That is, we always implicitly assume that the Γ-translates of Q cover $\mathbf{R}^n$ with overlaps of measure zero. ◇

Equation (2.8) applied to the attractor $Q = K_D$ has the form

$$Q = K_D = \sum_{j=1}^{\infty} A^{-j}(D) = \Big\{\sum_{j=1}^{\infty} A^{-j}\varepsilon_j : \varepsilon_j \in D\Big\}. \tag{2.9}$$

Thus, each point $x \in Q$ can be written $x = \sum_{j=1}^{\infty} A^{-j}\varepsilon_j$ for some $\varepsilon_j \in D$. We write $x = .\varepsilon_1\varepsilon_2\cdots$ in this case, and refer to this representation of x as an *A-nary expansion* of x. Note that A-nary expansions need not be unique.

EXAMPLE 2.5. Let $n = 1$, $\Gamma = \mathbf{Z}$, $A = 2$, and $\Lambda = \{0, \dots, N\}$ (allowing the possibility that $c_k = 0$ for some $k \in \Lambda$). In this case, the refinement equation has the form $f(x) = \sum_{k=0}^{N} c_k\, f(2x - k)$.

We have $m = |\det(A)| = 2$, and the sublattice $A(\Gamma)$ is the set of even integers $2\mathbf{Z}$. There are two cosets, $2\mathbf{Z}$ and $2\mathbf{Z} + 1$. We choose $D = \{0, 1\}$ as our full set of digits. The affine maps w_k defined by (2.5) are $w_k(x) = \frac{1}{2}(x + k)$ for $k \in \mathbf{Z}$. The attractor $Q = K_D$ is defined by the requirement that (2.7) hold, which translates to the statement that $Q = \frac{1}{2}Q \cup \frac{1}{2}(Q + 1)$. This is satisfied for the compact set $Q = [0, 1]$. Since $\{[0, 1] + k\}_{k \in \mathbf{Z}}$ covers $\mathbf{R}$ with overlaps of measure zero, this attractor Q is indeed a tile. Moreover, equation (2.9) states that each $x \in [0, 1]$ can be written $x = \sum_{j=1}^{\infty} 2^{-j} \varepsilon_j$ with $\varepsilon_j \in D = \{0, 1\}$, which is the binary expansion of x. ◊

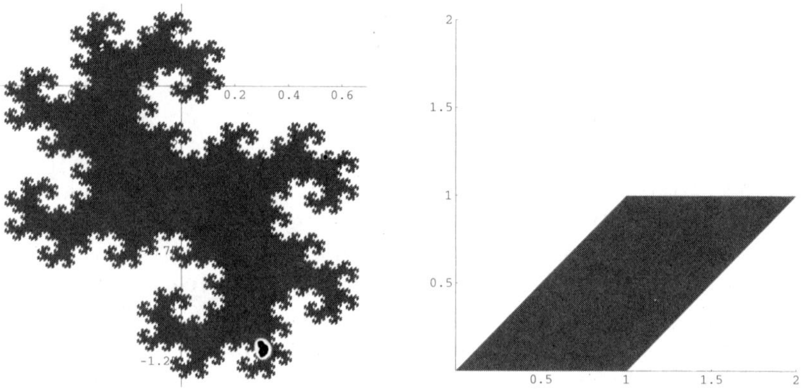

FIGURE 2.1. Twin Dragon and Parallelogram Attractors.

EXAMPLE 2.6. The tile Q may have a fractal boundary. For example, if $A_1 = \left[\begin{smallmatrix} 1 & -1 \\ 1 & 1 \end{smallmatrix}\right]$ and $D = \{(0,0),\, (1,0)\}$, then the tile Q is the celebrated "twin dragon" fractal shown on the left in Figure 2.1. On the other hand, if we choose $A_2 = \left[\begin{smallmatrix} 1 & 1 \\ 1 & -1 \end{smallmatrix}\right]$ and $D = \{(0,0),\, (1,0)\}$, then the tile Q is the parallelogram with vertices $\{(0,0),\, (1,0),\, (2,1),\, (1,1)\}$ pictured on the right in Figure 2.1. For these two matrices A_1 and A_2, the sublattices $A_1(\mathbf{Z}^2)$ and $A_2(\mathbf{Z}^2)$ coincide. This sublattice is called the *quincunx sublattice* of $\mathbf{Z}^2$. As a consequence, these two matrices A_1, A_2 are often referred to as *quincunx dilation matrices.* ◊

2.2.3. Covering by Translates of Q. We saw in Proposition 2.2 that if f is a compactly supported solution of the refinement equation (1.1), then $\operatorname{supp}(f) \subset K_\Lambda$. Since K_Λ is compact and Q is a tile, there exists a finite set $\Omega \subset \Gamma$ such that

$$K_\Lambda \subset Q + \Omega,$$

where

$$Q + \Omega = \bigcup_{\omega \in \Omega} (Q + \omega) = \{q + \omega : q \in Q,\, \omega \in \Omega\}.$$

If the tile Q is fractal-like, it may be difficult to construct such a set Ω. The next proposition gives one explicit example of a finite Ω with this property.

PROPOSITION 2.7. *Define*

$$\Lambda' = \Lambda - D = \{k - d : k \in \Lambda,\, d \in D\},$$

and let $K_{\Lambda'}$ *be the attractor of the IFS generated by* $\{w_k\}_{k\in\Lambda'}$. *Then* $\Omega_{\Lambda'} = K_{\Lambda'} \cap \Gamma$ *satisfies* $K_\Lambda \subset Q + \Omega_{\Lambda'}$. *Further,* $(Q+k) \cap K_\Lambda \neq \emptyset$ *for each* $k \in \Omega_{\Lambda'}$.

PROOF. Fix any $x \in K_\Lambda$. Since Q is a tile, we can write $x = q + k$ for some $q \in Q$ and $k \in \Gamma$. By (2.8) applied to $x \in K_\Lambda$ and $q \in Q = K_D$, we can write $x = \sum_{j=1}^\infty A^{-j}\lambda_j$ and $q = \sum_{j=1}^\infty A^{-j}\varepsilon_j$ with $\lambda_j \in \Lambda$ and $\varepsilon_j \in D$. Therefore,

$$k = x - q = \sum_{j=1}^\infty A^{-j}(\lambda_j - \varepsilon_j) \in K_{\Lambda'} \cap \Gamma = \Omega_{\Lambda'}.$$

Hence $x = q + k \in Q + \Omega_{\Lambda'}$.

Finally, suppose that $k \in \Omega_{\Lambda'}$, say $k = \sum_{j=1}^\infty A^{-j}(\lambda_j - \varepsilon_j)$. Then $x = \sum_{j=1}^\infty A^{-j}\lambda_j \in K_\Lambda$ and $q = \sum_{j=1}^\infty A^{-j}\varepsilon_j \in Q$, so $k + q = x \in (Q+k) \cap K_\Lambda$, and therefore $(Q+k) \cap K_\Lambda \neq \emptyset$. □

Thus, translates of the tile Q by elements of $\Omega_{\Lambda'}$ cover K_Λ, and hence the support of f. Moreover, $\Omega_{\Lambda'}$ is minimal in the sense that each of the translates $Q + k$ for $k \in \Omega_{\Lambda'}$ will intersect K_Λ, although it is possible that many of these intersections may have measure zero. It is often the case that smaller sets Ω can be found which also have the property that $K_\Lambda \subset Q + \Omega$. In particular, this is the case in the one-dimensional setting and also for the examples we present in Chapter 5.

EXAMPLE 2.8. Note that in the 1-D case, if $\Lambda = \{0, \dots, N\}$ then the attractor K_Λ of the IFS generated by $\{w_k\}_{k\in\Lambda}$ is the interval $K_\Lambda = [0, N]$, and therefore the scaling function f must be supported in this interval. The set Λ' defined in Proposition 2.7 is $\Lambda' = \Lambda - D = \{-1, \dots, N\}$. Then $K_{\Lambda'} = [-1, N]$ and $\Omega_{\Lambda'} = \{-1, \dots, N\} = \Lambda'$, so $K_\Lambda = [0, N] \subset [-1, N+1] = [0,1] + \{-1, \dots, N\} = Q + \Omega_{\Lambda'}$, in accordance with Proposition 2.7. However, the smaller set $\Omega = \{0, \dots, N-1\}$ also has the property that $K_\Lambda \subset Q + \Omega$. Indeed, $Q + \Omega = [0,1] + \{0, \dots, N-1\} = [0, N] = K_\Lambda$ in this case. ◇

If Ω is a finite subset of Γ such that $K_\Lambda \subset Q + \Omega$, and if $y \subset K_\Lambda$, then $y = x + k$ for some $x \in Q$ and $k \in \Omega$. However, it might also be the case that $y = x' + k'$ with $x' \in Q$ and $k' \notin \Omega$. The following lemma shows that this is impossible if y lies in the interior K_Λ° of K_Λ.

LEMMA 2.9. *Let* Ω *be a finite subset of* Γ. *If* $x + k \in (Q+\Omega)^\circ$ *with* $x \in Q$ *and* $k \in \Gamma$, *then* $k \in \Omega$. *In particular, if* $K_\Lambda \subset Q + \Omega$ *and* $x + k \in K_\Lambda^\circ$ *with* $x \in Q$ *and* $k \in \Gamma$, *then* $k \in \Omega$.

PROOF. Let $x \in Q$ and $k \in \Gamma$, and suppose that $x+k \in (Q+\Omega)^\circ$. Then we can find an open ball $B(x+k, \varepsilon)$ entirely contained in $(Q+\Omega)^\circ$. Define $F = B(x, \varepsilon) \cap Q$. By Lemma 2.3(b), the tile Q is the closure of its interior, so F must have positive Lebesgue measure, i.e., $|F| > 0$. If $y = z + k \in F + k$, then $|(x+k) - y| = |x - z| < \varepsilon$, so

$$F + k \subset B(x+k, \varepsilon) \subset (Q+\Omega)^\circ \subset Q + \Omega.$$

However, $F \subset Q$, so

$$F + k \subset (Q+k) \cap (Q+\Omega) = \bigcup_{j \in \Omega} (Q+k) \cap (Q+j).$$

If $k \notin \Omega$, then $|(Q+k) \cap (Q+j)| = 0$ by Lemma 2.3(c), which contradicts the fact that $|F| > 0$. Therefore we must have $k \in \Omega$. □

By our Standing Assumption, the Γ-translates of Q cover $\mathbf{R}^n$ with overlaps of measure zero (in fact, by Lemma 2.3, the overlaps will occur only on the boundaries of the translates of Q). We next prove that Q can be modified so that it tiles without overlaps. This is analogous to removing one endpoint from the interval $[0,1]$ so that integer translates of the resulting interval $[0,1)$ cover $\mathbf{R}$ without overlaps.

PROPOSITION 2.10. *Assume that Q is a tile. Then there exists $\tilde{Q} \subset Q$, such that the Γ-translates of $\tilde{Q}$ cover $\mathbf{R}^n$ without overlaps, i.e.,*

$$\tilde{Q} + \Gamma = \mathbf{R}^n \qquad \text{and} \qquad \tilde{Q} \cap (\tilde{Q}+k) = \emptyset \text{ for } k \in \Gamma \setminus \{0\}.$$

Further, $\tilde{Q} \cap \Gamma$ contains a single element.

PROOF. Divide the lattice Γ into disjoint subsets Γ^+, Γ^-, and $\{0\}$ in such a way that $\Gamma^- = -\Gamma^+$ and both Γ^+ and Γ^- are closed under vector addition. Specifically, let

$$\Gamma^+ = \bigcup_{i=1}^{n} \{k \in \Gamma : k = (k_1, \dots, k_i, 0, \dots, 0),\ k_i > 0\} \tag{2.10}$$

and let $\Gamma^- = -\Gamma^+$. Define

$$\tilde{Q} = Q \setminus \bigcup_{k \in \Gamma^+} (Q+k).$$

First we prove that the Γ-translates of $\tilde{Q}$ are disjoint. Suppose that we had $x \in \tilde{Q} \cap (\tilde{Q}+k)$ for some $k \in \Gamma^+$. Then since $x \in \tilde{Q}$, we have $x \in Q$ but $x \notin Q + j$ for any $j \in \Gamma^+$, which contradicts the fact that $x \in \tilde{Q} + k$. On the other hand, if $x \in \tilde{Q} \cap (\tilde{Q}+k)$ for some $k \in \Gamma^-$ then $x - k \in \tilde{Q} \cap (\tilde{Q} + (-k))$ with $(-k) \in \Gamma^+$, which reduces to the previous case. Since $\Gamma^+ \cup \Gamma^- = \Gamma \setminus \{0\}$, we conclude that Γ-translates of $\tilde{Q}$ are indeed disjoint.

Now we show that $\tilde{Q} + \Gamma = \mathbf{R}^n$. Since $Q + \Gamma = \mathbf{R}^n$, it suffices to show that $Q \subset \tilde{Q} + \Gamma$. So, suppose that $x \in Q$ but $x \notin \tilde{Q} + \Gamma$. Then we cannot have $x \in \tilde{Q}$, so we must have $x \in Q \setminus \tilde{Q}$. Therefore, by definition of $\tilde{Q}$, there exists a $j_1 \in \Gamma^+$ such that $x - j_1 \in Q$. If $x - j_1 \in \tilde{Q}$ then we would have $x \in \tilde{Q} + j_1 \subset \tilde{Q} + \Gamma$, which is a contradiction. Hence $x - j_1 \in Q \setminus \tilde{Q}$. Since we also clearly have $x - j_1 \notin \tilde{Q} + \Gamma$, we can repeat this argument to obtain a sequence of points $j_i \in \Gamma^+$ such that $x - \sum_{i=1}^{\ell} j_i \in Q \setminus \tilde{Q}$ for each ℓ. However, it is easy to see from the definition of Γ^+ that $\left\|\sum_{i=1}^{\ell} j_i\right\| \to \infty$, so this contradicts the fact that Q is compact.

Finally, since the Γ-translates of $\tilde{Q}$ do not overlap and cover all of $\mathbf{R}^n$, there must be a unique element of Γ that lies in $\tilde{Q}$. □

REMARK 2.11. (a) Proposition 2.10 remains valid if the specific sets Γ^+ and Γ^- defined by (2.10) are replaced by arbitrary subsets of Γ which have the properties that $\Gamma = \Gamma^+ \cup \Gamma^- \cup \{0\}$ disjointly, Γ^+ and Γ^- are closed under vector addition, and $\Gamma^- = -\Gamma^+$.

(b) Since we assume that 0 is one of the digits, the tile Q will contain 0. However, while $\tilde{Q}$ will contain a unique element of Γ, that element need not be 0. For example, if $n = 2$, $A = 2I$, and $D = \{(0,0), (1,0), (0,-1), (1,-1)\}$, then $Q = [0,1] \times [-1,0]$ and $\tilde{Q} = [0,1) \times [-1,0)$. ♢

2.3. Matrix Form of the Refinement Operator

Suppose that $f\colon \mathbf{R} \to \mathbf{C}$ is a compactly supported solution of the one-dimensional, single-function refinement equation

$$f(x) = \sum_{k=0}^{N} c_k\, f(2x-k), \qquad x \in \mathbf{R}. \tag{2.11}$$

Then f must be supported in the interval $[0,N]$. Further, the refinement equation can be recast into a matrix-vector form as follows. Define a vector-valued function $\Phi f\colon [0,1] \to \mathbf{C}^N$ by

$$\Phi f(x) = [f(x+k)]_{k=0}^{N-1} = \begin{bmatrix} f(x) \\ f(x+1) \\ \vdots \\ f(x+N-1) \end{bmatrix}, \qquad x \in [0,1]. \tag{2.12}$$

Since $\operatorname{supp}(f) \subset [0,N]$, the information in Φf is "equivalent" to the information in f. Define two matrices

$$T_0 = [c_{2j-k}]_{j,k=0}^{N-1} = \begin{bmatrix} c_0 & 0 & 0 & \cdots & 0 & 0 \\ c_2 & c_1 & c_0 & \cdots & 0 & 0 \\ \vdots & \vdots & \vdots & \ddots & \vdots & \vdots \\ 0 & 0 & 0 & \cdots & c_N & c_{N-1} \end{bmatrix} \tag{2.13}$$

and

$$T_1 = [c_{2j-k+1}]_{j,k=0}^{N-1} = \begin{bmatrix} c_1 & c_0 & 0 & \cdots & 0 & 0 \\ c_3 & c_2 & c_1 & \cdots & 0 & 0 \\ \vdots & \vdots & \vdots & \ddots & \vdots & \vdots \\ 0 & 0 & 0 & \cdots & 0 & c_N \end{bmatrix}. \tag{2.14}$$

Then the refinement equation (2.11) is equivalent to the equation

$$\Phi f(x) = \begin{cases} T_0\, \Phi f(2x), & 0 \le x \le 1/2, \\ T_1\, \Phi f(2x-1), & 1/2 < x \le 1. \end{cases} \tag{2.15}$$

Note that by using the $2x \bmod 1$ map given by

$$2x \bmod 1 = \begin{cases} 2x, & 0 \le x \le 1/2, \\ 2x-1, & 1/2 < x \le 1, \end{cases}$$

we can rewrite (2.15) as

$$\Phi f(x) = T_{d_1} \Phi f(2x \bmod 1), \tag{2.16}$$

where $x = .d_1 d_2 \dots$ is the binary expansion of x (for $x = 1/2$, use $1/2 = .1000\dots$).

We next define a function on $\mathbf{R}^n$ that is analogous to the $2x \bmod 1$ map and use it to obtain a matrix form of the general refinement equation that is analogous to (2.16).

DEFINITION 2.12. By definition, $Q = \bigcup_{i=1}^m w_{d_i}(Q)$. If $x \in Q$ is such that $x \in w_{d_i}(Q)$ for a *unique* digit d_i, then we set

$$\tau x \;=\; w_{d_i}^{-1}(x) \;=\; Ax - d_i. \tag{2.17}$$

Thus, if $x = .\varepsilon_1\varepsilon_2\cdots$ is an A-nary expansion of such an x, then $\varepsilon_1 = d_i$ and $\tau x = .\varepsilon_2\varepsilon_3\cdots$. For other x, the meaning of (2.1) is ambiguous. We eliminate this ambiguity by "disjointizing" the sets $w_{d_i}(Q)$. Specifically, we define

$$Q_1 \;=\; w_{d_1}(Q) \quad \text{and} \quad Q_i \;=\; w_{d_i}(Q) \setminus \Bigl(\bigcup_{j<i} Q_j\Bigr) \qquad \text{for } i = 2, \dots, m. \tag{2.18}$$

Then $Q_i \subset w_{d_i}(Q)$, and Q is the union of the disjoint sets $Q_1, \dots, Q_m$. Hence each $x \in Q$ lies in a unique Q_i, and we define τx by (2.17) using that unique value of i. $\diamondsuit$

Now let $\Omega \subset \Gamma$ be any fixed finite set such that

$$K_\Lambda \;\subset\; Q + \Omega.$$

For example, the set $\Omega_{\Lambda'}$ constructed in Proposition 2.7 is one possibility for Ω.

Given a function $g\colon \mathbf{R}^n \to \mathbf{C}^r$ with $\operatorname{supp}(g) \subset K_\Lambda$, we define the *folding* of g to be the function $\Phi g\colon Q \to (\mathbf{C}^{r\times 1})^{\Omega\times 1}$ given by

$$\Phi g(x) \;=\; [g(x+k)]_{k\in\Omega}, \qquad x \in Q.$$

If we write $(\Phi g)_k(x) = g(x+k)$ for the kth component of $\Phi g(x)$, then this folding has the property that $(\Phi g)_{k_1}(x_1) = (\Phi g)_{k_2}(x_2)$ whenever $x_1, x_2 \in Q$ and $k_1, k_2 \in \Omega$ are such that $x_1 + k_1 = x_2 + k_2$ (by Lemma 2.3(c), such points x_1, x_2 would have to lie on ∂Q).

For each d in our digit set D, define an $\Omega \times \Omega$ matrix T_d by

$$T_d \;=\; [c_{Aj-k+d}]_{j,k\in\Omega}. \tag{2.19}$$

Let $Q_1, \dots, Q_m$ be defined as in (2.18). Define an operator T acting on vector functions

$$u(x) = [u_k(x)]_{k\in\Omega}\colon Q \to (\mathbf{C}^{r\times 1})^{\Omega\times 1}$$

by

$$Tu(x) \;=\; \sum_{i=1}^{m} \chi_{Q_i}(x) \cdot T_{d_i} u(Ax - d_i). \tag{2.20}$$

Or, equivalently, T can be defined by

$$Tu(x) \;=\; T_{d_i} u(\tau x) \quad \text{if } x \in Q_i. \tag{2.21}$$

This operator T is related to the refinement operator S as follows.

PROPOSITION 2.13. *Let $\Omega \subset \Gamma$ be such that $K_\Lambda \subset Q + \Omega$. Assume that $g\colon \mathbf{R}^n \to \mathbf{C}^r$ satisfies*

$$\operatorname{supp}(g) \subset K_\Lambda \qquad \textit{and} \qquad g(x) = 0 \textit{ for } x \in \partial K_\Lambda.$$

(a) *If $x \in Q$ and $d \in D$ is such that $Ax - d \in Q$, then*

$$\Phi S g(x) = T_d \Phi g(Ax - d). \tag{2.22}$$

(b) $\Phi S g \;=\; T \Phi g.$

PROOF. (a) Let $x \in Q$, and let $y = Ax - d \in Q$. Suppose $g(y+k) \neq 0$ for some $k \in \Gamma$. Then $y + k \in K_\Lambda^\circ$, and therefore $k \in \Omega$ by Lemma 2.9. Hence,

$$\begin{aligned}
\Phi S g(x) &= [Sg(x+j)]_{j\in\Omega} \\
&= \left[\sum_{k\in\Gamma} c_k\, g(Ax - d + Aj - k + d)\right]_{j\in\Omega} \\
&= \left[\sum_{k\in\Gamma} c_{Aj-k+d}\, g(y+k)\right]_{j\in\Omega} \\
&= \left[\sum_{k\in\Omega} c_{Aj-k+d}\, g(y+k)\right]_{j\in\Omega} \\
&= T_d \Phi g(y) \\
&= T_d \Phi g(Ax - d). \qquad (2.23)
\end{aligned}$$

(b) Let $x \in Q$, and let $d = d_i$, where i is the unique integer such that $x \in Q_i$. Then $\tau x = Ax - d_i \in Q$, so by (2.22) and (2.21) we have that $\Phi S g(x) = T_{d_i} \Phi g(\tau x) = T \Phi g(x)$, and this is valid for every $x \in Q$. □

REMARK 2.14. Note that equation (2.22) is more general than the statement $\Phi S g(x) = T \Phi g(x)$. In particular, (2.22) reduces to the statement that $\Phi S g(x) = T_d \Phi g(\tau x) = T \Phi g(x)$ if it is the case that $d = d_i$, where i is the unique integer such that $x \in Q_i$. However, (2.22) is valid given only that $Ax - d \in Q$, and we will need to use this more general statement later. ◇

The equality in Proposition 2.13(b) is a pointwise everywhere equality. We show next that if we instead require only equality almost everywhere then the hypothesis in Proposition 2.13 that $g(x)$ vanish on the boundary of K_Λ can be removed.

COROLLARY 2.15. *Let $\Omega \subset \Gamma$ be such that $K_\Lambda \subset Q + \Omega$. If $g\colon \mathbf{R}^n \to \mathbf{C}^r$ satisfies* $\operatorname{supp}(g) \subset K_\Lambda$, *then*

$$\Phi S g = T \Phi g \quad a.e.$$

PROOF. Define $\tilde{g}(x) = g(x)$ for $x \in K_\Lambda^\circ$ and $\tilde{g}(x) = 0$ otherwise. Since ∂K_Λ has measure zero by Lemma 2.1(e), we have $g = \tilde{g}$ a.e. Proposition 2.13 therefore implies that $\Phi S \tilde{g} = T \Phi \tilde{g}$ pointwise everywhere. Since $Sg = S\tilde{g}$ a.e. and $T\Phi g = T \Phi \tilde{g}$ a.e., the result follows. □

EXAMPLE 2.16. Consider the one-dimensional refinement equation (2.11), but allow the multiplicity r to be arbitrary. We have $\operatorname{supp}(f) \subset K_\Lambda = [0, N]$. Hence $\Omega = \{0, \dots, N-1\}$ is the smallest subset of $\Gamma = \mathbf{Z}$ which has the property that $K_\Lambda \subset Q + \Omega$. With this choice of Ω, the folding of f is $\Phi f(x) = [f(x+k)]_{k=0}^{N-1}$, which coincides with (2.12) except that the entries $f(x+k)$ are now column vectors of length r. The digit set is $D = \{0, 1\}$, so there are two matrices $T_0 = [c_{2j-k}]_{j,k=0}^{N-1}$ and $T_1 = [c_{2j-k+1}]_{j,k=0}^{N-1}$. These coincide with the definitions in (2.13) and (2.14) except that the entries c_k are now $r \times r$ blocks. Finally, the recasting of the refinement equation performed in Corollary 2.15 reduces exactly to (2.16), except that the multiplicity r is now arbitrary. ◇

2.4. The Joint Spectral Radius

The spectral radius of a square matrix M is

$$\rho(M) \;=\; \lim_{\ell\to\infty} \|M^\ell\|^{1/\ell} \;=\; \max\{|\lambda| : \lambda \text{ is an eigenvalue of } M\}.$$

For each $1 \le p \le \infty$, the *p-joint spectral radius* (p-JSR) of a finite collection of $s \times s$ matrices $\mathcal{M} = \{M_1, \dots, M_m\}$ is

$$\hat{\rho}_p(\mathcal{M}) \;=\; \begin{cases} \displaystyle\lim_{\ell\to\infty} \Bigl(\sum_{\Pi\in\mathcal{P}_\ell} \|\Pi\|^p\Bigr)^{1/p\ell}, & 1 \le p < \infty, \\ \displaystyle\lim_{\ell\to\infty} \max_{\Pi\in\mathcal{P}_\ell} \|\Pi\|^{1/\ell}, & p = \infty, \end{cases} \tag{2.24}$$

where

$$\mathcal{P}_0 \;=\; \{I\} \qquad \text{and} \qquad \mathcal{P}_\ell \;=\; \{M_{j_1} \cdots M_{j_\ell} : 1 \le j_i \le m\}.$$

It is easy to see that the limit in (2.24) exists and is independent of the choice of norm $\|\cdot\|$ on $\mathbf{C}^{s\times s}$. Note that if $p \ge q$, then $\hat{\rho}_p(\mathcal{M}) \le \hat{\rho}_q(\mathcal{M})$.

We will refer to the ∞-JSR as the *uniform joint spectral radius*; it is also known as the *generalized spectral radius*, or simply as the *joint spectral radius*. Berger and Wang [**BW92**] proved that $\hat{\rho}_\infty(\mathcal{M}) < 1$ if and only if every product $M_{j_1} \cdots M_{j_\ell}$ converges to the zero matrix as $\ell \to \infty$, and that

$$\hat{\rho}_\infty(\mathcal{M}) \;=\; \lim_{\ell\to\infty} \max_{\Pi\in\mathcal{P}_\ell} \rho(\Pi)^{1/\ell}. \tag{2.25}$$

The proof of (2.25) is nontrivial when $\mathcal{M}$ contains more than one matrix. It follows from (2.25) that

$$\hat{\rho}_\infty(\mathcal{M}) \;=\; \sup\{|\lambda|^{1/\ell} : \ell > 0 \text{ and } \lambda \text{ is an eigenvalue of some } \Pi \in \mathcal{P}_\ell\}. \tag{2.26}$$

Note that if there is a norm such that $\bigl(\sum_{j=1}^m \|M_j\|^p\bigr)^{1/p} \le \delta$, then, by the definition of $\hat{\rho}$, it is clear that $\hat{\rho}_p(\mathcal{M}) \le \delta$. We next prove the following partial converse to this fact.

PROPOSITION 2.17. *Assume that $\mathcal{M} = \{M_1, \dots, M_m\}$ is a finite collection of $s \times s$ matrices. If $\hat{\rho}_p(\mathcal{M}) < \delta$, then there exists a vector norm $\|\cdot\|$ on $\mathbf{C}^s$ such that:*

(a) $\displaystyle\Bigl(\sum_{j=1}^m \|M_j x\|^p\Bigr)^{1/p} \le \delta\, \|x\|$ *for each $x \in \mathbf{C}^s$, if $1 \le p < \infty$, or*

(b) $\displaystyle\max_j \|M_j\| \le \delta$, *if $p = \infty$.*

PROOF. Assume first that $1 \le p < \infty$. Let $|\cdot|$ be any vector norm on $\mathbf{C}^s$, and define $\hat{\rho}_{p,\ell} = \bigl(\sum_{\Pi\in\mathcal{P}_\ell} |\Pi|^p\bigr)^{1/p\ell}$. Choose any number θ such that $\hat{\rho}_p(\mathcal{M}) < \theta < \delta$. Then since $\hat{\rho}_{p,\ell} \to \hat{\rho}_p(\mathcal{M})$, there must be some m such that $\hat{\rho}_{p,m} \le \theta$. Given any ℓ, write $\ell = mk + r$ with $0 \le r \le m - 1$. Define

$$C \;=\; \max\bigl\{(\hat{\rho}_{p,m})^{-i}\,(\hat{\rho}_{p,i})^i \;:\; i = 0, \dots, m-1\bigr\}.$$

Then

$$\begin{aligned}
(\hat{\rho}_{p,\ell})^{p\ell} &= \sum_{\Pi \in \mathcal{P}_\ell} |\Pi|^p \\
&= \sum_{\Pi_1 \in \mathcal{P}_m} \cdots \sum_{\Pi_k \in \mathcal{P}_m} \sum_{\Pi_0 \in \mathcal{P}_r} |\Pi_1 \cdots \Pi_k \Pi_0|^p \\
&\leq \Bigl(\sum_{\Pi_1 \in \mathcal{P}_m} |\Pi_1|^p\Bigr) \cdots \Bigl(\sum_{\Pi_k \in \mathcal{P}_m} |\Pi_k|^p\Bigr)\Bigl(\sum_{\Pi_0 \in \mathcal{P}_r} |\Pi_0|^p\Bigr) \\
&= (\hat{\rho}_{p,m})^{pkm} \, (\hat{\rho}_{p,r})^{pr} \\
&= (\hat{\rho}_{p,m})^{p\ell} \, (\hat{\rho}_{p,m})^{-pr} \, (\hat{\rho}_{p,r})^{pr} \\
&\leq C^p \, \theta^{p\ell}.
\end{aligned}$$

Therefore, for each $x \in \mathbf{C}^s$ and each $\ell \geq 0$ we have

$$\frac{1}{\theta^{p\ell}} \sum_{\Pi \in \mathcal{P}_\ell} |\Pi x|^p \;\leq\; \frac{|x|^p}{\theta^{p\ell}} \sum_{\Pi \in \mathcal{P}_\ell} |\Pi|^p \;\leq\; C^p \, |x|^p. \tag{2.27}$$

Let $\alpha > 1$ be that number such that $\alpha^{1/p}\,\theta = \delta$. Then the fact that $\alpha > 1$, combined with (2.27), implies that the series in the following definition converges for each $x \in \mathbf{C}^s$:

$$\|x\| \;=\; \biggl(\sum_{\ell=0}^{\infty} \frac{1}{\alpha^\ell} \frac{1}{\theta^{p\ell}} \sum_{\Pi \in \mathcal{P}_\ell} |\Pi x|^p\biggr)^{1/p}. \tag{2.28}$$

It is easy to verify that $\|\cdot\|$ defined by (2.28) is a vector norm on $\mathbf{C}^s$, and that $|x| \leq \|x\| \leq \bigl(\frac{\alpha}{\alpha-1}\bigr)^{1/p} |x|$. Finally, for each $x \in \mathbf{C}^s$ we have

$$\begin{aligned}
\sum_{j=1}^{m} \|M_j x\|^p &= \sum_{j=1}^{m} \sum_{\ell=0}^{\infty} \frac{1}{\alpha^\ell} \frac{1}{\theta^{p\ell}} \sum_{\Pi \in \mathcal{P}_\ell} |\Pi M_j x|^p \\
&= \sum_{\ell=0}^{\infty} \frac{1}{\alpha^\ell} \frac{1}{\theta^{p\ell}} \sum_{j=1}^{m} \sum_{\Pi \in \mathcal{P}_\ell} |\Pi M_j x|^p \\
&= \sum_{\ell=0}^{\infty} \frac{\alpha}{\alpha^{\ell+1}} \frac{\theta^p}{\theta^{p(\ell+1)}} \sum_{\Pi \in \mathcal{P}_{\ell+1}} |\Pi x|^p \\
&\leq \alpha \, \theta^p \, \|x\|^p \\
&= \delta^p \, \|x\|^p.
\end{aligned}$$

This completes the proof for the case $1 \leq p < \infty$. The proof for the case $p = \infty$ is similar, using the norm

$$\|x\| \;=\; \sup_{\ell \geq 0} \max_{\Pi \in \mathcal{P}_\ell} \frac{|\Pi x|}{\delta^\ell}$$

in place of (2.28). □

REMARK 2.18. We briefly illustrate why the joint spectral radius arises naturally in connection with refinement equations. Suppose that $f\colon \mathbf{R} \to \mathbf{C}$ is a continuous, compactly supported solution of the refinement equation $f(x) = \sum_{k=0}^{N} c_k\, f(2x-k)$. Then by (2.16), we have for each $x \in [0,1]$ that $\Phi f(x) = T_{\varepsilon_1} \Phi f(\tau x)$, where $x = .d_1 d_2 \dots$ is the binary representation of x.

Suppose now that $x_\ell = .\varepsilon_1\varepsilon_2 \cdots \varepsilon_\ell \varepsilon_{\ell+1} \varepsilon_{\ell+2} \cdots$ and $y_\ell = .\varepsilon_1\varepsilon_2 \cdots \varepsilon_\ell \varepsilon'_{\ell+1} \varepsilon'_{\ell+2} \cdots$ are points whose binary expansions agree for the first ℓ digits. Then we can iterate (2.16) to obtain

$$\begin{aligned}\Phi f(x_\ell) - \Phi f(y_\ell) &= T_{\varepsilon_1}\big(\Phi f(\tau x_\ell) - \Phi f(\tau y_\ell)\big) \\ &= T_{\varepsilon_1} T_{\varepsilon_2}\big(\Phi f(\tau^2 x_\ell) - \Phi f(\tau^2 y_\ell)\big) \\ &= \cdots = T_{\varepsilon_1} \cdots T_{\varepsilon_\ell}\big(\Phi f(\tau^\ell x_\ell) - \Phi f(\tau^\ell y_\ell)\big).\end{aligned}$$

As ℓ increases, the points x_ℓ and y_ℓ grow closer together. Since f and hence Φf is continuous, the difference $\Phi f(x_\ell) - \Phi f(y_\ell)$ must converge to the zero vector as $\ell \to \infty$. Since $\tau^\ell x_\ell$ and $\tau^\ell y_\ell$ can be arbitrary points in $[0,1]$, it follows that the product $T_{\varepsilon_1} \cdots T_{\varepsilon_\ell}$ must converge to zero as ℓ increases, at least when applied to vectors in the subspace

$$W_0 \;=\; \operatorname{span}\{\Phi f(w) - \Phi f(z) : w, z \in [0,1]\},$$

which can be shown to be a common invariant subspace for both T_0 and T_1. Therefore, a necessary condition for the existence of a continuous solution to the refinement equation is that all products $(T_{\varepsilon_1} \cdots T_{\varepsilon_\ell})|_{W_0}$ of T_0 and T_1 restricted to this invariant subspace W_0 must converge to zero as $\ell \to \infty$. By [**BW92**], this occurs if and only if $\hat\rho_\infty(\{T_0|_{W_0}, T_1|_{W_0}\}) < 1$. The space W_0 as given above is defined only implicitly, and is usually difficult or impossible in practice to determine explicitly, whereas in Theorem 3.4 and 3.21 we determine explicit subspaces to use in place of W_0 to characterize the L^p and continuous solutions of the refinement equation. $\diamondsuit$

CHAPTER 3

Generalized Self-Similarity and the Refinement Equation

3.1. Generalized Self-Similarity

A subset B of a set X is said to be *self-similar* if there exist injective maps $w_1, \dots, w_m \colon X \to X$ such that

$$B = \bigcup_{i=1}^{m} w_i(B).$$

Let X and H be sets. A function $f \colon X \to H$ is self-similar if its graph is self-similar, i.e.,

$$f(x) = f(w_i^{-1}(x)), \qquad x \in w_i(X), \; i = 1, \dots, m.$$

We say that $f \colon X \to H$ is a *generalized self-similar function* if there exist functions $\varphi_i \colon X \times H \to H$ and a function $\mathcal{O} \colon X \times H^m \to H$ such that

$$f(x) = \mathcal{O}(x, \varphi_1(x, f(w_1^{-1}(x))), \dots, \varphi_m(x, f(w_m^{-1}(x)))), \qquad x \in X.$$

The theory of generalized self-similar functions was developed in [**CM99**].

The following theorem is a variation on the results of [**CM99**]

THEOREM 3.1. *Let $1 \le p \le \infty$ be given. Let X be a compact subset of $\mathbf{R}^n$, and let H be a closed subset of $\mathbf{C}^r$. Let $\|\cdot\|$ be any norm on $\mathbf{C}^r$. Let $m \ge 1$, and assume that functions w_i, φ_i, and $\mathcal{O}$ are chosen with the following properties.*

1. *For each $i = 1, \dots, m$, let $w_i \colon X \to X$ be continuously differentiable, injective maps.*
2. *Let $\varphi_i \colon X \times H \to H$ for $i = 1, \dots, m$ satisfy a Lipschitz condition in the second variable, i.e.,*

$$\left(\sum_{i=1}^{m} \|\varphi_i(x, u) - \varphi_i(x, v)\|^p \right)^{1/p} \le C\, \|u - v\|, \tag{3.1}$$

   with the usual modification if $p = \infty$.

3. *Let $\mathcal{O} \colon X \times H^m \to H$ be non-expansive for each $x \in X$, i.e.,*

$$\|\mathcal{O}(x, u_1, \dots, u_m) - \mathcal{O}(x, v_1, \dots, v_m)\| \le \left(\sum_{i=1}^{m} \|u_i - v_i\|^p \right)^{1/p}, \tag{3.2}$$

   with the usual modification if $p = \infty$.

Let t_0 be an arbitrary point in H. For $u \in L^p(X, H)$, define

$$Tu(x) = \mathcal{O}(x, \varphi_1(x, u(w_1^{-1}(x))), \dots, \varphi_m(x, u(w_m^{-1}(x)))), \tag{3.3}$$

where we interpret

$$u(w_i^{-1}(x)) \;=\; t_0 \quad \textit{if } x \notin w_i(X). \tag{3.4}$$

Define

$$s \;=\; \max_{1\le i\le m}\ \sup_{x\in X} |\det((\mathrm{Diff}\, w_i)(x))|, \tag{3.5}$$

where Diff *is the differential operator. If* $\mathcal{O}$ *and the* φ_i *map bounded sets into bounded sets, then* T *maps* $L^p(X,H)$ *into itself, and satisfies*

$$\|Tu - Tv\|_{L^p} \;\le\; s^{1/p}\, C\, \|u-v\|_{L^p}. \tag{3.6}$$

In particular, if $s^{1/p}\, C < 1$*, then* T *is contractive, and there exists a unique function* $v^* \in L^p(X,H)$ *such that* $Tv^* = v^*$*. Moreover, in this case, if* $v^{(0)}$ *is any function in* $L^p(X,H)$*, then the iteration* $v^{(i+1)} = Tv^{(i)}$ *converges to* v^* *in* $L^p(X,H)$*.*

PROOF. The fact that T maps $L^p(X,H)$ into itself can be proved using the same techniques as in [**CM99**]. Therefore we will only prove that T satisfies the Lipschitz condition in (3.6). Given u, $v \in L^p(X,H)$, we have that

$$\begin{aligned}
&\|Tu - Tv\|_{L^p}^p \\
&= \int_X \|\mathcal{O}(x, \varphi_1(x, u(w_1^{-1}(x))), \dots, \varphi_m(x, u(w_m^{-1}(x)))) - \\
&\qquad\qquad \mathcal{O}(x, \varphi_1(x, v(w_1^{-1}(x))), \dots, \varphi_m(x, v(w_m^{-1}(x))))\|^p \, dx \\
&\le \int_X \sum_{i=1}^m \|\varphi_i(x, u(w_i^{-1}(x))) - \varphi_i(x, v(w_i^{-1}(x)))\|^p \, dx && \text{by (3.2)} \\
&= \sum_{i=1}^m \int_X \|\varphi_i(x, u(w_i^{-1}(x))) - \varphi_i(x, v(w_i^{-1}(x)))\|^p \, dx \\
&= \sum_{i=1}^m \int_{w_i(X)} \|\varphi_i(x, u(w_i^{-1}(x))) - \varphi_i(x, v(w_i^{-1}(x)))\|^p \, dx && \text{by (3.4)} \\
&\le s \sum_{i=1}^m \int_X \|\varphi_i(w_i(x), u(x)) - \varphi_i(w_i(x), v(x))\|^p \, dx && \text{by (3.5)} \\
&\le s\, C^p \int_X \|u(x) - v(x)\|^p \, dx && \text{by (3.1)} \\
&= s\, C^p \, \|u-v\|_{L^p}^p. \quad \square
\end{aligned}$$

3.2. Sufficient Conditions for the Existence of Vector Scaling Functions

The *accuracy* of a refinable vector function or distribution f is the largest integer $\kappa > 0$ such that every multivariate polynomial $q(x) = q(x_1, \dots, x_n)$ with $\deg(q) < \kappa$ can be written

$$q(x) \;=\; \sum_{k\in\Gamma} a_k f(x+k) \;=\; \sum_{k\in\Gamma}\sum_{i=1}^{r} a_{k,i} f_i(x+k) \text{ a.e.}, \qquad x \in \mathbf{R}^n,$$

for some row vectors $a_k = (a_{k,1}, \dots, a_{k,r}) \in \mathbf{C}^{1\times r}$. If no polynomials are reproducible from translates of f then we set $\kappa = 0$. We say that translates of f along Γ are *linearly independent* if $\sum_{k\in\Gamma} a_k f(x+k) = 0$ implies $a_k = 0$ for each k.

For the main result of this section (Theorem 3.4), we will need to impose only the minimal accuracy condition $\kappa = 1$. The following lemma from [**CHM98**] characterizes minimal accuracy.

LEMMA 3.2. *Let f be a compactly supported distributional solution of the refinement equation* (1.1). *Let $\Gamma_d = A(\Gamma) - d$ denote the cosets defined in* (2.2).

(a) *If there exists a row vector $u_0 \in C^{1\times r}$ such that $u_0 \hat{f}(0) \neq 0$ and*

$$u_0 \;=\; \sum_{k\in\Gamma_d} u_0 c_k \quad \text{for each } d \in D, \tag{3.7}$$

then f has accuracy $\kappa \geq 1$, and

$$\sum_{k\in\Gamma} u_0 f(x+k) \;=\; 1 \text{ a.e.} \tag{3.8}$$

(b) *If f has accuracy $\kappa \geq 1$ and if f has independent translates, then there exists a row vector $u_0 \in C^{1\times r}$ such that $u_0 \hat{f}(0) \neq 0$ and* (3.7) *holds.*

The hypothesis of linear independence of translates in Lemma 3.2(b) can be weakened.

In the single-function setting ($r = 1$), equation (3.7) reduces to the requirement that $\sum_{k\in\Gamma_d} c_k = 1$ for each $d \in D$.

Note that if (3.7) holds then, since the m cosets Γ_d for $d \in D$ partition Γ, u_0 is a left 1-eigenvector for the matrix $\Delta = \frac{1}{m}\sum c_k$. If this eigenvalue is nondegenerate and if the remaining eigenvalues are less than 1 in absolute value, then the following proposition from [**CHM00**] implies that a distributional solution f to the refinement equation does exist.

PROPOSITION 3.3. *If the matrix $\Delta = \frac{1}{m}\sum_{k\in\Lambda} c_k$ has eigenvalues $\lambda_1 = \cdots = \lambda_s = 1$ and $|\lambda_{s+1}|, \dots, |\lambda_r| < 1$ with the eigenvalue* 1 *nondegenerate, then there exist compactly supported distributions $f_1, \dots, f_r$ such that $f = (f_1, \dots, f_r)^T$ satisfies the refinement equation* (1.1) *in the sense of distributions. Furthermore, $\hat{f}(\omega)$ is a continuous vector function, and $\hat{f}(0) \neq 0$.*

To motivate the following result, suppose that f is a continuous, compactly supported vector scaling function with accuracy $\kappa \geq 1$. By Lemma 2.2, we have $\operatorname{supp}(f) \subset K_\Lambda$. Let Ω be any finite subset of Γ such that $K_\Lambda \subset Q + \Omega$. Let u_0 be the row vector such that (3.8) holds, i.e., $\sum_{k\in\Gamma} u_0 f(x+k) = 1$ a.e. If $x \in Q$, then Lemma 2.9 implies that the only nonzero terms in this series occur when $k \in \Omega$. Hence, if we set $e_0 = (u_0)_{k\in\Omega}$, i.e., e_0 is the row vector in $(\mathbf{C}^{1\times r})^{1\times\Omega}$ obtained by repeating the block u_0 once for each $k \in \Omega$, then

$$e_0 \Phi f(x) \;=\; \sum_{k\in\Omega} u_0 f(x+k) \;=\; \sum_{k\in\Gamma} u_0 f(x+k) \;=\; 1 \text{ a.e.}, \qquad x \in Q.$$

Thus the values of $\Phi f(x)$ are constrained to lie in a particular hyperplane H in $(\mathbf{C}^{r\times 1})^{\Omega\times 1}$, namely, the collection of vectors $v = [v_k]_{k\in\Omega}$ such that $e_0 v = \sum_{k\in\Omega} u_0 v_k = 1$. Further, the set of differences $E_0 = H - H$ is the subspace consisting of vectors $v = [v_k]_{k\in\Omega}$ such that $e_0 v = \sum_{k\in\Omega} u_0 v_k = 0$. Define the dot product of two column vectors $u = [u_k]_{k\in\Omega}$ and $v = [v_k]_{k\in\Omega} \in (\mathbf{C}^{r\times 1})^{\Omega\times 1}$ by

$$u \cdot v \;=\; u^* v \;=\; \sum_{k\in\Omega} u_k^* v_k \;=\; \sum_{k\in\Omega}\sum_{i=1}^{r} \bar{u}_{k,i} v_{k,i},$$

where u^* is the Hermitian, or conjugate transpose, of u. Then $e_0 v = e_0^* \cdot v$, so E_0 is simply the orthogonal complement of the column vector e_0^*.

The following theorem gives conditions for the existence of a continuous or L^p vector scaling function under the assumption of minimal accuracy.

THEOREM 3.4. *Let $1 \le p \le \infty$ be fixed. Let $\Omega \subset \Gamma$ be a finite set such that $K_\Lambda \subset Q + \Omega$. Assume that there exists a nonzero vector $u_0 \in \mathbf{C}^{1\times r}$ such that $u_0 = \sum_{k\in\Gamma_d} u_0 c_k$ for every $d \in D$, cf. equation (3.7). Let $e_0 = (u_0)_{k\in\Omega} \in (\mathbf{C}^{1\times r})^{1\times\Omega}$, and define*

$$E_0 \;=\; (e_0^*)^\perp \;=\; \Big\{v = [v_k]_{k\in\Omega} : e_0 v = \sum_{k\in\Omega} u_0 v_k = 0\Big\}. \tag{3.9}$$

Set

$$I_0^p \;=\; \Big\{g \in L^p(\mathbf{R}^n, \mathbf{C}^r) \;:\; \operatorname{supp}(g) \subset K_\Lambda \;\; \textit{and} \;\; \sum_{k\in\Gamma} u_0 g(x+k) = 1 \text{ a.e.}\Big\}. \tag{3.10}$$

If

$$I_0^p \neq \emptyset \qquad \textit{and} \qquad \hat{\rho}_p(\{T_d|_{E_0}\}_{d\in D}) \;<\; m^{1/p},$$

then there exists a unique function $f \in I_0^p$ which is a solution to the refinement equation (1.1), and the cascade algorithm $f^{(i+1)} = Sf^{(i)}$ converges geometrically in L^p-norm to f for each $f^{(0)} \in I_0^p$. Furthermore, if $p = \infty$ and I_0^∞ contains a continuous function, then f is continuous.

PROOF. Define

$$H \;=\; \{v = [v_k]_{k\in\Omega} \in (\mathbf{C}^{r\times 1})^{\Omega\times 1} : e_0 v = \sum_{k\in\Omega} u_0 v_k = 1\}. \tag{3.11}$$

It follows from (3.7) that e_0 is a common left 1-eigenvector for each matrix T_d, so if $e_0 v = 1$, then $e_0(T_d v) = (e_0 T_d) v = e_0 v = 1$. Thus H is right-invariant under each T_d. Further, the set E_0 given by (3.9) satisfies $E_0 = H - H$, is a subspace of $(\mathbf{C}^{r\times 1})^{\Omega\times 1}$, and is right-invariant under each matrix T_d.

Assume that $1 \le p < \infty$. We will apply Theorem 3.1 with $X = Q$ and H as given by (3.11). Our first step is to define functions w_d. φ_d, and $\mathcal{O}$ that satisfy the hypotheses of Theorem 3.1.

For $d \in D$, define $w_d(x) = A^{-1}(x + d)$. Then clearly each w_d is injective and continuously differentiable. Further, $\det((\operatorname{Diff} w_d)(x)) = 1/m$ for every x.

Let δ be any number such that

$$\hat{\rho}_p(\{T_d|_{E_0}\}_{d\in D}) \;<\; \delta \;<\; m^{1/p}.$$

Then by Proposition 2.17 applied to the matrices $T_d|_{E_0}$, there exists a vector norm $\|\cdot\|_{E_0}$ on E_0 such that

$$\sum_{d\in D} \|T_d w\|_{E_0}^p \;\le\; \delta^p \, \|w\|_{E_0}^p, \qquad \text{all } w \in E_0.$$

Let $\|\cdot\|$ denote any extension of this norm to all of $(\mathbf{C}^{r\times 1})^{\Omega\times 1}$.

Since $T_d(H) \subset H$, we can define $\varphi_d\colon Q \times H \to H$ by $\varphi_d(x,u) = T_d u$. Then, since $H - H \subset E_0$, we have for each $x \in Q$ and $u, v \in H$ that

$$\sum_{d\in D} \|\varphi_d(x,u) - \varphi_d(x,v)\|^p = \sum_{d\in D} \|T_d(u-v)\|^p \leq \delta^p \|u-v\|^p.$$

Therefore the functions φ_d satisfy the condition (3.1) with constant $C = \delta$. It is easy to check that each φ_d maps bounded sets into bounded sets.

Let $Q_1, \dots, Q_m$ be the disjoint subsets of Q defined by (2.18), and define $\mathcal{O}\colon Q \times H^m \to H$ by

$$\mathcal{O}(x, u_1, \dots, u_m) = \sum_{i=1}^{m} \chi_{Q_i}(x) \cdot u_i.$$

That is, $\mathcal{O}(x, u_1, \dots, u_m) = u_i$ if $x \in Q_i$. Then $\mathcal{O}$ maps bounded sets to bounded sets and satisfies the nonexpansivity condition (3.2).

Now let T be defined by (3.3), i.e., for $u \in L^p(Q,H)$ define

$$Tu(x) = \sum_{i=1}^{m} \chi_{Q_i}(x) \cdot T_{d_i} u(Ax - d_i).$$

That is, $Tu(x) = T_{d_i} u(\tau x)$ if $x \in Q_i$. Note that this operator T coincides with the operator T defined in (2.20). Since the number s defined by (3.5) has the value $s = 1/m$, Theorem 3.1 implies that T maps $L^p(Q,H)$ into itself, and satisfies

$$\|Tu - Tv\|_{L^p} \leq m^{-1/p}\, \delta\, \|u - v\|_{L^p}.$$

Since $\delta < m^{1/p}$, it follows that T is contractive on $L^p(Q,H)$ and there exists a unique function $v^* \in L^p(Q,H)$ such that $Tv^* = v^*$. Further, the iteration $v^{(i+1)} = Tv^{(i)}$ converges geometrically in $L^p(Q,H)$ to v^* for each function $v^{(0)} \in L^p(Q,H)$.

Clearly I_0^p is a closed subset of $L^p(\mathbf{R}^n, \mathbf{C}^r)$, and we claim that it is invariant under the refinement operator S. To see this, suppose that $g \in I_0^p$. First, we clearly have $Sg \in L^p(\mathbf{R}^n, \mathbf{C}^r)$ since $g \in L^p(\mathbf{R}^n, \mathbf{C}^r)$ and Λ is finite. Second, since $\operatorname{supp}(g) \subset K_\Lambda$, we have $\operatorname{supp}(Sg) \subset K_\Lambda$ by Proposition 2.2. Finally, to complete the claim we must show that $\sum_{k\in\Gamma} u_0 Sg(x+k) = 1$ a.e. Suppose that $x \in Q^\circ$ and $k \in \Gamma$ is such that $x + k \in \operatorname{supp}(g)$. Then we have $x + k \in \operatorname{supp}(g) \subset K_\Lambda \subset Q + \Omega$. However, the fact that x lies in the interior of Q combined with the fact that lattice translates of Q intersect only on the boundaries of these translates implies that $x + k \in (Q + \Omega)^\circ$. Lemma 2.9 therefore implies that $k \in \Omega$. Since this is valid for every $x \in Q^\circ$ and since ∂Q has measure zero, we conclude that

$$e_0 \Phi g(x) = \sum_{k\in\Omega} u_0 g(x+k) = \sum_{k\in\Gamma} u_0 g(x+k) = 1, \qquad \text{a.e. } x \in Q.$$

Thus $\Phi g(x) \in H$ for a.e. $x \in Q$. By Corollary 2.15, $\Phi S g = T \Phi g$. Since H is invariant under each matrix T_d, we therefore have $\Phi Sg(x) \in H$ for a.e. $x \in Q$. Since $\operatorname{supp}(Sg)$ is also included in K_Λ, we can again apply Lemma 2.9 to conclude that

$$1 = e_0 \Phi Sg(x) = \sum_{k\in\Omega} u_0 Sg(x+k) = \sum_{k\in\Gamma} u_0 Sg(x+k), \qquad \text{a.e. } x \in Q.$$

Since Q tiles $\mathbf{R}^n$ by translates along Γ, we conclude that this equality actually holds for a.e. $x \in \mathbf{R}^n$. Thus $Sg \in I_0^p$, so I_0^p is invariant under S as claimed.

In summary, the statements above combined with Corollary 2.15 imply that the following diagram commutes, with T in particular being a contraction:

$$\begin{array}{ccc} I_0^p & \xrightarrow{\Phi} & L^p(Q,H) \\ {\scriptstyle S}\downarrow & & \downarrow{\scriptstyle T} \\ I_0^p & \xrightarrow[\Phi]{} & L^p(Q,H). \end{array}$$

Suppose that $f^{(0)}$ is any function in I_0^p, and define $f^{(i+1)} = Sf^{(i)}$. Then $f^{(i)} \in I_0^p$ for each i. If we set $v^{(i)} = \Phi f^{(i)}$, then

$$v^{(i+1)} \;=\; \Phi f^{(i+1)} \;=\; \Phi S f^{(i)} \;=\; T\Phi f^{(i)} \;=\; Tv^{(i)},$$

so $v^{(i)}$ must converge in L^p-norm to v^*.

We now choose some particular norms for these L^p spaces. Let $|\cdot|$ be any fixed norm on $\mathbf{C}^r$. Then

$$\|g\|_{L^p}^p \;=\; \int_{\mathbf{R}^n} |g(x)|^p\,dx, \qquad g \in L^p(\mathbf{R}^n, \mathbf{C}^r),$$

defines an equivalent norm for $L^p(\mathbf{R}^n, \mathbf{C}^r)$. Similarly,

$$\|G\|_{L^p}^p \;=\; \int_Q \|G(x)\|^p\,dx, \qquad G \in L^p(Q, (\mathbf{C}^{r\times 1})^{\Omega\times 1}),$$

defines an equivalent norm for $L^p(Q, (\mathbf{C}^{r\times 1})^{\Omega\times 1})$.

Now define a norm $|\!|\!|\cdot|\!|\!|$ on $(\mathbf{C}^{r\times 1})^{\Omega\times 1}$ by

$$|\!|\!|w|\!|\!| \;=\; \left(\sum_{k\in\Omega} |w_k|^p\right)^{1/p}, \qquad w = [w_k]_{k\in\Omega} \in (\mathbf{C}^{r\times 1})^{\Omega\times 1}.$$

Since all norms on a finite-dimensional space are equivalent, we can find a constant $B > 0$ such that $|\!|\!|\cdot|\!|\!| \le B\,\|\cdot\|$. Therefore, if $g \in L^p(\mathbf{R}^n, \mathbf{C}^r)$ is supported in K_Λ, then since $K_\Lambda \subset Q + \Omega$ we have

$$\begin{aligned} \|g\|_{L^p}^p &= \int_{Q+\Omega} |g(x)|^p\,dx, \\ &= \sum_{k\in\Omega} \int_Q |g(x+k)|^p\,dx \\ &= \int_Q |\!|\!|\Phi g(x)|\!|\!|^p\,dx \\ &\le B^p \int_Q \|\Phi g(x)\|^p\,dx \\ &= B^p\,\|\Phi g\|_{L^p}^p. \end{aligned}$$

In particular,

$$\|f^{(i)} - f^{(j)}\|_{L^p} \;\le\; B\,\|\Phi f^{(i)} - \Phi f^{(j)}\|_{L^p} \;=\; B\,\|v^{(i)} - v^{(j)}\|_{L^p},$$

so $f^{(i)}$ must converge in L^p-norm to some function $f \in L^p(\mathbf{R}^n, \mathbf{C}^r)$. We must have $f \in I_0^p$ since I_0^p is a closed subset of $L^p(\mathbf{R}^n, \mathbf{C}^r)$. Further,

$$\Phi f \;=\; v^* \;=\; Tv^* \;=\; T\Phi f \;=\; \Phi S f \text{ a.e.},$$

the last equality following from Corollary 2.15. Therefore f satisfies the refinement equation (1.1) almost everywhere. Since v^* is unique, the cascade algorithm must converge to this particular f for any starting function $f^{(0)} \in I_0^p$.

This completes the proof for the case $1 \le p < \infty$. The argument to this point for the case $p = \infty$ is entirely similar. It therefore only remains observe that if any $f^{(0)} \in I_0^\infty$ is continuous, then the iterates $f^{(i)}$ obtained from $f^{(0)}$ are continuous and converge uniformly to f, so f must itself be continuous. $\square$

EXAMPLE 3.5. Consider the one-dimensional setting with $A = 2$, $\Gamma = \mathbf{Z}$, $D = \{0,1\}$, $\Lambda = \{0,\dots,N\}$, $K_\Lambda = [0,N]$, and $\Omega = \{0,\dots,N-1\}$. There are two cosets, $\Gamma_0 = 2\mathbf{Z}$ and $\Gamma_1 = 2\mathbf{Z}+1$, so the minimal accuracy condition (3.7) reduces to the requirement that there exists a row vector $u_0 \in \mathbf{C}^{1\times r}$ such that

$$u_0 \;=\; \sum_{k\in\mathbf{Z}} u_0 c_{2k} \;=\; \sum_{k\in\mathbf{Z}} u_0 c_{2k+1}.$$

The row vector e_0 is formed by repeating the row vector u_0 once for each $k \in \Omega = \{0,\dots,N-1\}$, i.e.,

$$e_0 \;=\; (u_0)_{k=0}^{N-1} \;=\; (u_0,\dots,u_0) \;\in\; (\mathbf{C}^{1\times r})^{1\times N}.$$

Hence E_0 consists of all the column vectors $v = (v_0,\dots,v_{N-1})^{\mathrm{T}} \in (\mathbf{C}^{r\times 1})^{N\times 1}$ such that

$$e_0 v \;=\; \sum_{k=0}^{N-1} u_0 v_k \;=\; 0.$$

Further, I_0^p consists of those L^p vector functions $g\colon \mathbf{R} \to \mathbf{C}^r$ which are supported in $[0,N]$ and which have the property that $\sum u_0 g(x+k) = 1$ a.e. In particular, if $N \ge 2$ and we let h be the hat function on $[0,2]$, i.e.,

$$h(x) \;=\; \max\{1 - |1-x|, 0\},$$

and let $a \in \mathbf{C}^{r\times 1}$ be a column vector satisfying $u_0 a = 1$, then I_0^p will contain the continuous function $g(x) = a h(x)$. Therefore, if $\hat\rho_p(T_0|_{E_0}, T_1|_{E_0}) < 2^{1/p}$, then there exists an L^p solution f to the refinement equation, and the cascade algorithm converges in L^p-norm to f for any starting function $f^{(0)}$ chosen from I_0^p. Further, if $p = \infty$, then f is continuous.

There are further simplifications in the single-function case ($r = 1$). In particular, if $r = 1$ then u_0 is a scalar, and by normalizing we can simply let $u_0 = 1$. $\diamondsuit$

The same techniques used to prove Theorem 3.4 can also be used to prove the following more general result.

THEOREM 3.6. *Let $1 \le p \le \infty$ be fixed. Let $\Omega \subset \Gamma$ be a finite set such that $K_\Lambda \subset Q+\Omega$. Let H be a nonempty closed subset of $(\mathbf{C}^{r\times 1})^{\Omega\times 1}$ such that $T_d(H) \subset H$ for each $d \in D$. Let E be a subspace of $(\mathbf{C}^{r\times 1})^{\Omega\times 1}$ which contains $H - H$ and which is right-invariant under each T_d. Define*

$$I_0^p \;=\; \Big\{ g \in L^p(\mathbf{R}^n, \mathbf{C}^r) \;:\; \operatorname{supp}(g) \subset K_\Lambda \;\text{ and }\; \Phi g(Q) \subset H \Big\}.$$

If $I_0^p \ne \emptyset$ and $\hat\rho_p(\{T_d|_E\}_{d\in D}) < m^{1/p}$, then there exists a function $f \in I_0^p$ which is a solution to the refinement equation (1.1), and the cascade algorithm $f^{(i+1)} = Sf^{(i)}$ converges in L^p-norm to f for each function $f^{(0)} \in I_0^p$. Furthermore, if

$\hat{\rho}_\infty(\{T_d|_E\}_{d\in D}) < 1$ *and there exists any continuous function* $f^{(0)} \in I_0^\infty$, *then* f *is continuous.*

3.3. Continuous Solutions and the Support of the Refinement Equation Coefficients

The set I_0^∞ defined by (3.10) is determined by two quantities: the set Λ and the row vector u_0. The set Λ is the support of the set of coefficients c_k in the refinement equation; it is determined only by the location of the c_k and not their values. The vector u_0, on the other hand, is determined by the values of the c_k as well as their locations. In this section we will consider what requirements must be placed on Λ and u_0 so that I_0^∞ will contain a continuous function. We will see that, in fact, this is determined solely by Λ and not by u_0, i.e., only the location of the coefficients c_k is important for this question, and not their actual values.

Since any continuous function supported in K_Λ must be zero on the boundary of K_Λ, it is sufficient to study the question of when the set

$$I(\Lambda, u_0) = \Big\{g \in L^\infty(\mathbf{R}^n, \mathbf{C}^r) \ : \ g(x) = 0 \text{ for } x \notin K_\Lambda^\circ \text{ and } \sum_{k\in\Gamma} u_0 g(x+k) = 1\Big\}$$

contains a continuous function. Here the notation $I(\Lambda, u_0)$ is meant to emphasize the dependence on Λ and u_0. The following result shows that $I(\Lambda, u_0)$ contains a continuous function exactly when it contains any functions at all. Further, whether $I(\Lambda, u_0)$ is nonempty or not is independent of the value of u_0.

LEMMA 3.7. *Let* $\Lambda \subset \Gamma$ *be finite, and let* $u_0 \in C^{1\times r}$ *be nonzero. Then the following statements are equivalent.*

(a) $I(\Lambda, u_0) \neq \emptyset$.

(b) $I(\Lambda, u_0)$ *contains a continuous function.*

(c) $K_\Lambda^\circ + \Gamma = \mathbf{R}^n$, *i.e., lattice translates of* K_Λ° *cover* $\mathbf{R}^n$.

PROOF. (a) $\Rightarrow$ (c). Assume there exists a function $g \in I(\Lambda, u_0)$. Then since $\sum_{k\in\Gamma} u_0 g(x+k)$ never vanishes but $g(x+k) \neq 0$ only for $x + k \in K_\Lambda^\circ$, we must have $\bigcup_{k\in\Gamma}(K_\Lambda^\circ + k) = \mathbf{R}^n$.

(c) $\Rightarrow$ (b). Suppose that $\bigcup_{k\in\Gamma}(K_\Lambda^\circ + k) = \mathbf{R}^n$. Then K_Λ has nonempty interior, so there exist continuous scalar-valued functions $h\colon \mathbf{R}^n \to C$ supported in K_Λ such that $h(x) > 0$ for each $x \in K_\Lambda^\circ$. For example, $h(x) = \operatorname{dist}(x, (K_\Lambda^\circ)^{\mathrm{C}})$ has this property. Let $a \in \mathbf{C}^{r\times 1}$ be such that $u_0 a = 1$, and define $s(x) = \sum_{k\in\Gamma} h(x+k)$. Then s is a continuous, scalar-valued function which never vanishes, and therefore $g(x) = ah(x)/s(x)$ is a continuous vector function which lies in $I(\Lambda, u_0)$. □

Since the size of the set Λ will determine the size of the matrices T_d, and therefore the complexity of the computation of the JSR, it should be chosen to be as small as possible, while satisfying the requirements of Lemma 3.7. However, even "large" Λ may fail the necessary condition $K_\Lambda^\circ + \Gamma = \mathbf{R}^n$, as the following example shows.

EXAMPLE 3.8. Let $n = 2$, and consider the uniform dilation $A = 2I$. With $\Gamma = \mathbf{Z}^2$, a natural digit choice is $D = \{(0,0), (1,0), (0,1), (1,1)\}$. Let s be any positive integer, and define

$$\Lambda = \{0,1\} \times \{0, \dots, s\}.$$

Then $K_\Lambda = [0,1] \times [0,s]$, so $K_\Lambda^\circ = (0,1) \times (0,s)$. Hence $\mathbf{Z}^2$-translates of K_Λ° do not cover $\mathbf{R}^n$, so Lemma 3.7 implies that $I(\Lambda, u_0)$ is empty. $\diamondsuit$

A related question is whether, for a given choice of dilation matrix A and digit set D, there must exist *some* finite set $\Lambda \subset \Gamma$ such that $K_\Lambda^\circ + \Gamma = \mathbf{R}^n$. This will always be the case. For example, if $\Lambda = D + D$ then it follows from (2.7) that $K_\Lambda = K_D + K_D = Q + Q$. Since $Q^\circ + Q = \bigcup_{q \in Q} (Q^\circ + q)$ is open, we have that $Q^\circ + Q \subset (Q+Q)^\circ = K_\Lambda^\circ$. Since Q is a tile, we know by Lemma 2.3(b) that $Q^\circ \neq \emptyset$. Therefore K_Λ° contains some translate $q_0 + Q$ of Q, and therefore $K_\Lambda^\circ + \Gamma \supset (q_0 + Q) + \Gamma = \mathbf{R}^n$.

3.4. Higher-Order Accuracy

We saw in Theorem 3.4 that if the coefficients c_k of the refinement equation satisfy (3.7), the condition for minimal accuracy, then the space E_0 defined by (3.9) is right-invariant under each matrix T_d. We will show below that if the coefficients c_k satisfy the conditions for higher-order accuracy then E_0 is only the largest of a decreasing chain of common invariant subspaces

$$E_0 \supset E_1 \supset \cdots \supset E_{\kappa-1},$$

and that, as a consequence, the value of $\hat{\rho}_\infty(\{T_d|_{E_0}\}_{d \in D})$ is determined by the value of $\hat{\rho}_\infty(\{T_d|_{E_{\kappa-1}}\}_{d \in D})$. Moreover, these invariant spaces E_s are directly determined from the coefficients c_k via the accuracy conditions, which are a system of linear equations. Hence it is a simple matter to compute the matrices $T_d|_{E_{\kappa-1}}$.

We will use the standard multi-index notation, i.e., $x^\alpha = x_1^{\alpha_1} \cdots x_n^{\alpha_n}$ where $\alpha = (\alpha_1, \dots, \alpha_n)$ is an n-tuple of nonnegative integers and $x \in \mathbf{R}^n$. The degree of α is

$$|\alpha| = \alpha_1 + \cdots + \alpha_n.$$

The number of multi-indices α of a given degree s is

$$d_s = \binom{s+n-1}{n-1}.$$

In particular, $d_0 = 1$ and $d_1 = n$. If $n = 1$ then $d_s = 1$ for each s, and if $n = 2$ then $d_s = s + 1$ for each s. We write $\beta \leq \alpha$ if $\beta_i \leq \alpha_i$ for $i = 1, \dots, n$. We set

$$\binom{\alpha}{\beta} = \begin{cases} \binom{\alpha_1}{\beta_1} \cdots \binom{\alpha_n}{\beta_n}, & \beta \leq \alpha, \\ 0, & \text{otherwise.} \end{cases}$$

We shall often deal with matrix-valued functions

$$u = [u_{j,k}]_{j \in J, k \in K} : \mathbf{R}^n \to \mathbf{C}^{J \times K},$$

each of whose entries $u_{j,k} : \mathbf{R}^n \to \mathbf{C}$ is a polynomial. In this case, we refer to u as a *matrix of polynomials*, and we say that the *degree* of u is

$$\deg(u) = \max\{\deg(u_{j,k})\}_{j \in J, k \in K}.$$

The following lemma shows that the accuracy of any function supported in K_Λ is necessarily finite.

LEMMA 3.9. *Assume* $g\colon \mathbf{R}^n \to \mathbf{C}^r$ *satisfies* $\mathrm{supp}(g) \subset K_\Lambda$. *Let* $\Omega \subset \Gamma$ *be such that* $K_\Lambda \subset Q + \Omega$. *Then the accuracy* κ *of* g *is bounded by the requirement that*

$$\sum_{s=0}^{\kappa-1} d_s \;\le\; r \cdot \#\Omega,$$

PROOF. Assume that g has accuracy κ. Then for each multi-index α with $|\alpha| < \kappa$, there exist row vectors $y_\alpha(k) = (y_{\alpha,1}(k), \dots, y_{\alpha,r}(k))$ such that

$$x^\alpha \;=\; \sum_{k\in\Gamma} y_\alpha(k)\, g(x+k) \;=\; \sum_{k\in\Gamma}\sum_{i=1}^{r} y_{\alpha,i}(k)\, g_i(x+k).$$

Let $x \in Q^\circ$. If $x + k \in \mathrm{supp}(g)$, then we have $x + k \in K_\Lambda \subset Q + \Omega$. But since $x \in Q^\circ$, this can only happen if $x + k \in (Q + \Omega)^\circ$. Lemma 2.9 therefore implies that $k \in \Omega$. Hence, if we restrict our attention to the set Q°, we have

$$x^\alpha|_{Q^\circ} \;\in\; \mathrm{span}\{g_i(x+k)|_{Q^\circ}\}_{k\in\Omega,\, i=1,\dots,r}.$$

Since Q° is a nonempty open set, the polynomials x^α restricted to Q° are linearly independent. Hence the total number of such polynomials, which is $\sum_{s=0}^{\kappa-1} d_s$, can be at most the dimension of $\mathrm{span}\{g_i(x+k)|_{Q^\circ}\}_{k\in\Omega,\, i=1,\dots,r}$, which is bounded by $r \cdot \#\Omega$. □

We require some notation in order to discuss higher-order accuracy. Proofs of the facts given below can be found in [**CHM98**], [**CHM00**]. For a given degree $s \ge 0$, we collect the d_s monomials x^α of degree s together to form a column vector of monomials $X_{[s]}\colon \mathbf{R}^n \to \mathbf{C}^{d_s}$. Specifically, $X_{[s]}$ is defined by

$$X_{[s]}(x) \;=\; [x^\alpha]_{|\alpha|=s}, \qquad x \in \mathbf{R}^n.$$

The ordering of the multi-indices α of degree s is not important, as long as the same ordering is used throughout.

For each integer $0 \le t \le s$, define a matrix of polynomials $Q_{[s,t]}\colon \mathbf{R}^n \to \mathbf{C}^{d_s\times d_t}$ by

$$Q_{[s,t]}(y) \;=\; (-1)^{s-t}\left[\binom{\alpha}{\beta} y^{\alpha-\beta}\right]_{|\alpha|=s,|\beta|=t},$$

where we use the convention that $0^0 = 1$. In particular, $Q_{[s,s]}(y) = I$, the $d_s \times d_s$ identity matrix. Translation of $X_{[s]}(x)$ obeys the rule

$$X_{[s]}(x-y) \;=\; \sum_{t=0}^{s} Q_{[s,t]}(y)\, X_{[t]}(x). \tag{3.12}$$

Given any $n \times n$ matrix $B = [b_{i,j}]_{i,j=1,\dots,n}$ with scalar entries and given $s \ge 0$, let $B_{[s]} = [b^s_{\alpha,\beta}]_{|\alpha|=s,|\beta|=s}$ be the $d_s \times d_s$ matrix whose scalar entries $b^s_{\alpha,\beta}$ are defined by the equation

$$\sum_{|\beta|=s} b^s_{\alpha,\beta}\, x^\beta \;=\; (Bx)^\alpha \;=\; \prod_{i=1}^{n} (b_{i,1}x_1 + \cdots + b_{i,n}x_n)^{\alpha_i}.$$

Dilation of $X_{[s]}(x)$ by B obeys the rule

$$X_{[s]}(Bx) \;=\; B_{[s]}\, X_{[s]}(x). \tag{3.13}$$

The matrix $B_{[s]}$ has a number of surprising properties. For example, if $\lambda = (\lambda_1, \dots, \lambda_n)^{\mathrm{T}}$ is the vector of all eigenvalues of B, then $[\lambda^\alpha]_{|\alpha|=s}$ is the vector of all eigenvalues of $B_{[s]}$.

Given a collection

$$\{v_\alpha = (v_{\alpha,1}, \dots, v_{\alpha,r}) \in \mathbf{C}^{1\times r} : 0 \le |\alpha| < \kappa\}$$

of row vectors of length r, we shall associate a number of special matrices and functions. First, we group the v_α by degree to form $d_s \times 1$ column vectors $v_{[s]} \in (\mathbf{C}^{1\times r})^{d_s\times 1}$ with block entries that are the $1 \times r$ row vectors v_α, i.e.,

$$v_{[s]} = [v_\alpha]_{|\alpha|=s} = \begin{bmatrix} v_{\alpha_1,1} & \cdots & v_{\alpha_1,r} \\ \vdots & \ddots & \vdots \\ v_{\alpha_{d_s},1} & \cdots & v_{\alpha_{d_s},r} \end{bmatrix}.$$

Note that $v_{[0]} = [v_0] = v_0$. Later we will choose v_0 to coincide with the vector u_0 appearing in (3.7).

Second, for each α, we define a row vector of polynomials $y_\alpha : \mathbf{R}^n \to \mathbf{C}^{1\times r}$ by

$$y_\alpha(x) = \sum_{0\le\beta\le\alpha} (-1)^{|\alpha|-|\beta|} \binom{\alpha}{\beta} v_\beta\, x^{\alpha-\beta}. \tag{3.14}$$

Note that if $v_0 \ne 0$, then $\deg(y_\alpha) = |\alpha|$. We will see in Theorem 3.12 that, under appropriate conditions on the vectors v_α, the row vectors $y_\alpha(k)$ are precisely those vectors such that $\sum_{k\in\Gamma} y_\alpha(k)\, f(x+k) = x^\alpha$.

As with the vectors v_α, we collect the vectors of polynomials y_α by degree and arrange them as block entries in a column vector to form a matrix of polynomials $y_{[s]} : \mathbf{R}^n \to (\mathbf{C}^{1\times r})^{d_s\times 1}$, i.e.,

$$\begin{aligned} y_{[s]}(x) &= [y_\alpha(x)]_{|\alpha|=s} \\ &= \left[\sum_{t=0}^{s} \sum_{|\beta|=t} (-1)^{s-t} \binom{\alpha}{\beta} x^{\alpha-\beta}\, v_\beta \right]_{|\alpha|=s} \\ &= \sum_{t=0}^{s} Q_{[s,t]}(x)\, v_{[t]}. \end{aligned}$$

Finally, for each x we collect the blocks $y_{[s]}(x+k)$ into an infinite row vector to form a function $Y_{[s]} : \mathbf{R}^n \to \left((\mathbf{C}^{1\times r})^{d_s\times 1}\right)^{1\times\Gamma}$, i.e.,

$$Y_{[s]}(x) = \big(y_{[s]}(x+k)\big)_{k\in\Gamma}.$$

Note that $Y_{[s]}(0) = \big(y_{[s]}(k)\big)_{k\in\Gamma}$ is the row vector of evaluations of the matrix of polynomials $y_{[s]}$ at lattice points $k \in \Gamma$.

EXAMPLE 3.10. In the one-dimensional setting $n = 1$, there is a single polynomial x^s of degree s. Therefore $d_s = 1$ for every s, and the multi-index α that has degree s is simply the scalar $\alpha = s$. Thus $A_{[s]}$ is a scalar and $X_{[s]}$ and $Q_{[s,t]}$ are scalar-valued functions on $\mathbf{R}$. In particular, with $A = 2$ and $\Gamma = \mathbf{Z}$ we have

$$A_{[s]} = 2^s, \qquad X_{[s]}(x) = x^s, \qquad Q_{[s,t]}(y) = (-1)^{s-t} \binom{s}{t} y^{s-t}.$$

Hence (3.12) is nothing more than the binomial theorem, and (3.13) is the statement that $(2x)^s = 2^s x^s$. The vectors $v_\alpha = v_s$ are "ordinary" row vectors of length r. Further, there is only one v_α to "stack" to form $v_{[s]}$, so $v_{[s]} = v_s$. The functions $y_\alpha(x) = y_s(x)$ are row vector-valued, and $y_{[s]}(x)$ is a "stack" of $y_s(x)$ alone, so equals $y_s(x)$. Thus,

$$v_\alpha = v_s \in \mathbf{C}^{1\times r},$$

$$v_{[s]} = [v_\alpha]_{|\alpha|=s} = v_s \in \mathbf{C}^{1\times r},$$

$$y_s(x) = \sum_{t=0}^{s} (-1)^{s-t} \binom{s}{t} x^{s-t} v_t \quad \text{maps } \mathbf{R} \to \mathbf{C}^{1\times r},$$

$$y_{[s]}(x) = [y_\alpha(x)]_{|\alpha|=s} = y_s(x) \qquad \text{maps } \mathbf{R} \to \mathbf{C}^{1\times r},$$

$$Y_{[s]}(x) = \big(y_s(x+k)\big)_{k\in\mathbf{Z}} \qquad \text{maps } \mathbf{R} \to (\mathbf{C}^{1\times r})^{1\times\mathbf{Z}}.$$

In particular, $Y_{[s]}(x)$ is an infinite row vector whose entries are the $1\times r$ row vectors $y_s(x+k)$ with $k \in \mathbf{Z}$. Thus $Y_{[s]}(x)$ is simply an "ordinary" infinite row vector of the form

$$Y_{[s]}(x) \;=\; \big(\cdots,\, y_s(x-1),\, y_s(x),\, y_s(x+1),\, \cdots\big),$$

with blocks $y_s(x+k)$ that are ordinary $1 \times r$ row vectors. $\diamondsuit$

The following fact on the behavior of the matrix of polynomials $y_{[s]}$ under translation will be useful.

LEMMA 3.11. *Given a collection* $\{v_\alpha \in \mathbf{C}^{1\times r} : 0 \le |\alpha| < \kappa\}$ *of row vectors, let* $y_{[s]}(x)$ *and* $Y_{[s]}(x)$ *be as defined above. Then*

$$y_{[s]}(x+y) \;=\; \sum_{t=0}^{s} Q_{[s,t]}(y)\, y_{[t]}(x),$$

$$Y_{[s]}(x+y) \;=\; \sum_{t=0}^{s} Q_{[s,t]}(y)\, Y_{[t]}(x).$$

The following result provides sufficient conditions for a refinable distribution to have accuracy κ [**CHM98**], [**CHM00**]. These conditions are also necessary if f has independent translates.

THEOREM 3.12. *Assume that f is a compactly supported distributional solution of the refinement equation* (1.1). *Define* $L = [c_{Ai-j}]_{i,j\in\Gamma}$, *and consider the following statements.*

(I) *f has accuracy κ.*

(II) *There exists a collection of row vectors* $\{v_\alpha \in \mathbf{C}^{1\times r} : 0 \le |\alpha| < \kappa\}$ *such that*

(i) $v_0 \hat{f}(0) \ne 0$, *and*

(ii) $Y_{[s]}(0) = A_{[s]}\, Y_{[s]}(0)\, L$ *for* $0 \le s < \kappa$.

Then the following statements hold.

(a) *If translates of f along Γ are independent, then statement* (I) *implies statement* (II).

(b) *Statement* (II) *implies statement* (I). *Moreover, in this case, after scaling the vectors v_α by an appropriate constant, we have*

$$X_{[s]}(x) \;=\; \sum_{k\in\Gamma} y_{[s]}(k)\, f(x+k) \;=\; Y_{[s]}(0)\, F(x), \qquad 0 \le s < \kappa, \tag{3.15}$$

where $F(x) = [f(x+k)]_{k\in\Gamma}$.

Note that (3.15) says exactly that

$$x^\alpha \;=\; \sum_{k\in\Gamma} y_\alpha(k)\, f(x+k), \qquad 0 \le |\alpha| < \kappa.$$

By (3.14), the coefficients $y_\alpha(k)$ have the form of vectors of polynomials $y_\alpha(x)$ evaluated at lattice points $k \in \Gamma$.

REMARK 3.13. The vector v_0 is a left 1-eigenvector for the matrix $\Delta = \frac{1}{m}\sum c_k$, and $\hat{f}(0)$ is a right 1-eigenvector for this same matrix. In most applications, the matrix Δ is chosen so that 1 is a simple eigenvalue (in particular, this is a necessary condition for f to have linearly independent translates). In this case v_0 and $\hat{f}(0)$ are unique up to scale and automatically satisfy the condition $v_0\hat{f}(0) \ne 0$. In particular, in the single-function setting v_0 and $\hat{f}(0)$ are both nonzero scalars, so their product is automatically nonzero. ◊

REMARK 3.14. It is proved in [**CHM98**, Theorem 4.8] that the condition that

$$Y_{[s]}(0) = A_{[s]}\, Y_{[s]}(0)\, L, \qquad \text{for } 0 \le s < \kappa, \tag{3.16}$$

can be restated as:

$$y_{[s]}(\ell) \;=\; A_{[s]} \sum_{k\in\Gamma} y_{[s]}(k)\, c_{Ak-\ell}, \qquad \text{for } 0 \le s < \kappa \text{ and } \ell \in \Gamma, \tag{3.17}$$

and that the set of infinitely many conditions on the vectors v_α given by (3.17) is in fact equivalent to the following finite system of finite linear equations:

$$v_{[s]} \;=\; \sum_{k\in\Gamma_d} \sum_{t=0}^{s} Q_{[s,t]}(k)\, A_{[t]}\, v_{[t]}\, c_k, \qquad \text{for } 0 \le s < \kappa \text{ and } d \in D, \tag{3.18}$$

where $\Gamma_d = A(\Gamma) - d$. Note that this system is in block triangular form in the variables $v_{[t]}$. The coefficients $Q_{[s,t]}(k)$, $A_{[t]}$, and c_k are all known explicitly. It can be shown that in the single-function setting ($r = 1$), the system (3.18) is solvable if and only if

$$\sum_{k\in\Gamma} c_k = m \qquad \text{and} \qquad \sum_{k\in\Gamma_{d_1}} k^\alpha c_k \;=\; \cdots \;=\; \sum_{k\in\Gamma_{d_m}} k^\alpha c_k \quad \text{for } 0 \le |\alpha| < \kappa,$$

where $D = \{d_1, \dots, d_m\}$ is a listing of the digits in some order. Note that this system of equations is determined by the coefficients c_k and the sublattice $A(\Gamma)$, and does not depend directly on the dilation matrix A, in contrast to (3.16), (3.17), or (3.18). ◊

3.5. Invariant Subspaces

We will now show that the assumption of higher-order accuracy conditions on the coefficients c_k imposes considerable structure on the matrices T_d. Specifically, we will show that these matrices share common eigenvalues and invariant subspaces.

Assume that the sufficient conditions for accuracy κ given in Statement II of Theorem 3.12 are satisfied. In particular, $v_0 \neq 0$, and therefore the vector of polynomials y_α defined by (3.14) has degree $|\alpha|$. The finite row vectors

$$e_\alpha \;=\; (y_\alpha(k))_{k\in\Omega} \;\in\; (\mathbf{C}^{1\times r})^{1\times\Omega}, \qquad 0 \le |\alpha| < \kappa, \tag{3.19}$$

formed by restricting the infinite row vectors $(y_\alpha(k))_{k\in\Gamma}$ to components whose indices lie in Ω will play an important role, as will their spans

$$U_s \;=\; \operatorname{span}\{e_\alpha : 0 \le |\alpha| \le s\}.$$

EXAMPLE 3.15. In the one-dimensional setting $n = 1$ there is only one multi-index α for each degree s, namely $\alpha = s$, so $U_s = \operatorname{span}\{e_0, \dots, e_s\}$. The vector of polynomials $y_s\colon \mathbf{R}^n \to \mathbf{C}^{1\times r}$ defined by (3.14) has the form

$$y_s(x) \;=\; \sum_{t=0}^{s} (-1)^{s-t} \binom{s}{t} v_t\, x^{s-t},$$

and since $\Omega = \{0, \dots, N-1\}$, we have

$$e_s \;=\; (y_s(k))_{k\in\Omega} \;=\; (y_s(0), \dots, y_s(N-1)).$$

In particular, $e_0 = (v_0, \dots, v_0)$.

Now further restrict to the single-function setting $r = 1$. In this case, y_s is a *scalar*-valued polynomial of degree s, and e_s is the row vector of length N whose components are the evaluations of the polynomial y_s at the integers $k = 0, \dots, N-1$. In particular, after rescaling f by an appropriate constant, we can take $e_0 = (1, \dots, 1)$. Since $e_s = (y_s(0), \dots, y_s(N-1))$ and y_s is a polynomial of degree s, it follows that the space $U_s = \operatorname{span}\{e_0, \dots, e_s\}$ consists of the vectors of evaluations of all polynomials of degree at most s at the points $0, \dots, N-1$. That is, if we let

$$P_s \;=\; \{u\colon \mathbf{R} \to \mathbf{C} \;:\; u = 0 \text{ or } u \text{ is a polynomial with } \deg(u) \le s\},$$

then

$$U_s \;=\; \{(u(0), \dots, u(N-1)) : u \in P_s\}.$$

Thus, while $\{e_0, \dots, e_s\}$ is a natural basis for U_s in the context of the accuracy conditions presented in Section 3.4, another natural basis for U_s is $\{w_0, \dots, w_s\}$, where

$$w_t \;=\; (0^t, 1^t, \dots, (N-1)^t).$$

Indeed, this basis often implicitly appears in papers dealing with accuracy of scaling functions in the one-dimensional, single-function setting. To compare these two bases, note that $e_s = (y_s(0), \dots, y_s(N-1))$ where y_s that polynomial such that $\sum_{k\in\mathbf{Z}} y_s(k) f(x+k) = x^s$, while $w_s = (q_s(0), \dots, q_s(N-1))$ where q_s is the monomial $q_s(x) = x^s$. In this case $\sum_{k\in\mathbf{Z}} q_s(k) f(x+k)$ is a polynomial in x of degree s, but in general it is not the polynomial x^s.

However, while both $\{e_0, \dots, e_s\}$ and $\{w_0, \dots, w_s\}$ are natural bases for U_s in the single-function setting, only the basis $\{e_0, \dots, e_s\}$ has a direct generalization to the multi-function setting. This is because if $r > 1$ and we let

$$P_{s,r} = \{u \colon \mathbf{R} \to \mathbf{C}^{1\times r} : u = 0 \text{ or } u \text{ is a vector polynomial with } \deg(u) \le s\},$$

then

$$U_s \subsetneq \{(u(0), \dots, u(N-1)) : u \in P_{s,r}\}, \tag{3.20}$$

because U_s has dimension $s+1$ while the set on the right-hand side of (3.20) has dimension $r(s+1)$. In other words, when $r > 1$ the space U_s contains only *some* of the possible vectors of evaluations of polynomials of degree at most s. Hence, in the multi-function setting, the vectors e_α *must* be computed in order to compute the space U_s. Analogues of these remarks carry over to the higher-dimensional setting as well. Fortunately, once the accuracy conditions in (3.18) are solved, the vectors e_α are easily and immediately computable from (3.19) and (3.14). ♢

We observe next that the vectors e_α are linearly independent if a solution to the refinement equation does exist. We stipulate that whenever we consider a collection such as $\{e_\alpha : 0 \le |\alpha| < \kappa\}$, we assume that the vectors in this set are ordered from lowest degree to highest, with the ordering within degree fixed but unimportant.

LEMMA 3.16. *Assume that there exists a compactly supported distributional solution f to the refinement equation* (1.1), *and that* Statement (II) *of* Theorem 3.12 *holds. Then the vectors e_α defined in* (3.19) *are linearly independent.*

PROOF. Theorem 3.12 implies that $x^\alpha = \sum_{k\in\Gamma} y_\alpha(k)\, f(x+k)$. If $x \in Q^\circ$, then Lemma 2.9 implies that $x+k \in \operatorname{supp}(f)$ can hold only when $k \in \Omega$. Hence, if there exist scalars h_α such that $\sum_{0\le|\alpha|<\kappa} h_\alpha\, e_\alpha = 0$, then for $x \in Q^\circ$ we have

$$0 = \sum_{0\le|\alpha|<\kappa} h_\alpha\, e_\alpha\, \Phi f(x) = \sum_{0\le|\alpha|<\kappa} h_\alpha \sum_{k\in\Omega} y_\alpha(k)\, f(x+k) = \sum_{0\le|\alpha|<\kappa} h_\alpha\, x^\alpha.$$

Since Q° is nonempty, this implies $h_\alpha = 0$ for every α. □

The next theorem states that the assumption of the higher-order accuracy conditions given by the equivalent equations (3.16)–(3.18) implies that the matrices T_d share common invariant subspaces. This result is essentially a statement about the coefficients c_k in the refinement equation and does not require the assumption that a solution to the refinement equation exist. By Lemma 3.16, if a compactly supported solution does exist, even merely in the distributional sense, then the hypothesis in the following theorem that the vectors e_α are independent is redundant. Recall that B^* denotes the Hermitian, or conjugate transpose, of a matrix B.

THEOREM 3.17. *Let $\Omega \subset \Gamma$ be a finite set such that $K_\Lambda \subset Q + \Omega$. Assume that there exist row vectors $\{v_\alpha \in \mathbf{C}^{1\times r} : 0 \le |\alpha| < \kappa\}$ such that* (3.18) *holds. Let e_α be defined as in* (3.19), *and assume that the vectors $\{e_\alpha : 0 \le |\alpha| < \kappa\}$ are linearly independent. Define*

$$U_s = \operatorname{span}\{e_\alpha : 0 \le |\alpha| \le s\} \subset (\mathbf{C}^{1\times r})^{1\times\Omega}$$

and

$$E_s = \{e_\alpha^* : 0 \le |\alpha| \le s\}^\perp = \{v \in (\mathbf{C}^{r\times 1})^{\Omega\times 1} : e_\alpha v = 0 \textit{ for } 0 \le |\alpha| \le s\}.$$

Then the following statements hold.

(a) *U_s is left-invariant under T_d for each $d \in D$.*

(b) *E_s is right-invariant under T_d for each $d \in D$.*

(c) *Let $\{\tilde{e}_\alpha : 0 \le |\alpha| < \kappa\}$ be the result of applying the Gram–Schmidt orthogonalization procedure to the system $\{e_\alpha : 0 \le |\alpha| < \kappa\}$. Let $\mathcal{B}_E$ be any orthonormal basis for $E_{\kappa-1}$. Then*

$$\mathcal{B} \;=\; \{\tilde{e}_\alpha^* : 0 \le |\alpha| < \kappa\} \cup \mathcal{B}_E \tag{3.21}$$

is an orthonormal basis for $(\mathbf{C}^{r\times 1})^{\Omega\times 1}$, and the matrix for T_d in this basis has the block form

$$[T_d]_{\mathcal{B}} \;=\; \begin{bmatrix} B_0 & & & & 0 \\ & B_1 & & & \\ & & \ddots & & \\ & & & B_{\kappa-1} & \\ * & & & & C_d \end{bmatrix}, \tag{3.22}$$

where each B_s is a fixed $d_s \times d_s$ matrix whose Jordan form coincides with the Jordan form for $A_{[s]}^{-1}$, and where $C_d = [T_d|_{E_{\kappa-1}}]_{\mathcal{B}_E}$. In particular, B_0 is the scalar 1.

(d) *$\hat{\rho}_\infty(\{T_d|_{E_0}\}_{d\in D}) \;=\; \max\{\rho(A^{-1}),\ \hat{\rho}_\infty(\{T_d|_{E_{\kappa-1}}\}_{d\in D})\}$.*

PROOF. (a) Let P_Ω be the projection matrix defined by

$$P_\Omega = [\delta_{j,k} I_r]_{j\in\Gamma,k\in\Omega},$$

where I_r is the $r \times r$ identity matrix and $\delta_{j,k} = 1$ if $j = k$ and 0 otherwise. Then $e_\alpha = (y_\alpha(k))_{k\in\Gamma}\, P_\Omega$. Therefore, if we form the column vector $[e_\alpha]_{|\alpha|=s}$ whose components are the row vectors e_α, then we can write

$$[e_\alpha]_{|\alpha|=s} \;=\; (y_{[s]}(k))_{k\in\Gamma}\, P_\Omega \;=\; Y_{[s]}(0)\, P_\Omega. \tag{3.23}$$

Combining this with the fact that

$$P_\Omega\, T_d \;=\; [c_{Ak-\ell+d}]_{k\in\Gamma,\ell\in\Omega}, \tag{3.24}$$

we compute that

$$\begin{aligned}
[e_\alpha T_d]_{|\alpha|=s} &= Y_{[s]}(0)\, P_\Omega\, T_d && \text{by (3.23)} \\
&= (y_{[s]}(k))_{k\in\Gamma}\, [c_{Ak-\ell+d}]_{k\in\Gamma,\ell\in\Omega} && \text{by (3.24)} \\
&= \Big(\sum_{k\in\Gamma} y_{[s]}(k)\, c_{Ak-\ell+d}\Big)_{\ell\in\Omega} && \\
&= \big(A_{[s]}^{-1}\, y_{[s]}(\ell - d)\big)_{\ell\in\Omega} && \text{by (3.17)} \\
&= A_{[s]}^{-1}\, Y_{[s]}(-d)\, P_\Omega && \text{by definition of } Y_{[s]} \\
&= A_{[s]}^{-1} \sum_{t=0}^{s} Q_{[s,t]}(-d)\, Y_{[t]}(0)\, P_\Omega && \text{by Lemma 3.11} \\
&= A_{[s]}^{-1} \sum_{t=0}^{s} Q_{[s,t]}(-d)\, [e_\beta]_{|\beta|=t} && \text{by (3.23).}
\end{aligned}$$

Therefore $e_\alpha T_d \in \operatorname{span}\{e_\beta : |\beta| \le s\} = U_s$ for each $|\alpha| = s$, so U_s is left-invariant under T_d.

(b) Follows immediately from (a).

(c) Note that

$$\mathcal{B}' = \{e^*_\alpha : 0 \le |\alpha| < \kappa\} \cup \mathcal{B}_E$$

is a basis for $(\mathbf{C}^{r\times 1})^{\Omega\times 1}$, and that the basis $\mathcal{B}$ is obtained from $\mathcal{B}'$ via Gram–Schmidt. Using the computations from part (a) and the fact that $Q_{[s,s]}(y) = I$, we have for each $0 \le s < \kappa$ that

$$\begin{aligned}
[e_\alpha T_d]_{|\alpha|=s} &= A^{-1}_{[s]} \sum_{t=0}^{s} Q_{[s,t]}(-d)\, [e_\beta]_{|\beta|=t} \\
&= A^{-1}_{[s]}\, [e_\alpha]_{|\alpha|=s} + A^{-1}_{[s]} \sum_{t=0}^{s-1} Q_{[s,t]}(-d)\, [e_\beta]_{|\beta|=t}. \qquad (3.25)
\end{aligned}$$

As a consequence, the matrix for T^*_d in the basis $\mathcal{B}'$ has the form

$$[T^*_d]_{\mathcal{B}'} = \begin{bmatrix} (A^{-1}_{[0]})^* & & & * \\ & \ddots & & \\ & & (A^{-1}_{[\kappa-1]})^* & \\ 0 & & & \tilde{C}^*_d \end{bmatrix}$$

for some matrix $\tilde{C}_d$. Recall that the Gram–Schmidt orthogonalization of vectors $w_1, \dots, w_\ell$ preserves $\operatorname{span}\{w_1, \dots, w_k\}$ for each $k = 1, \dots, \ell$. Therefore,

$$[T^*_d]_{\mathcal{B}} = \begin{bmatrix} B^*_0 & & & * \\ & \ddots & & \\ & & B^*_{\kappa-1} & \\ 0 & & & C^*_d \end{bmatrix}$$

where B_s is a $d_s \times d_s$ matrix obtained from $A^{-1}_{[s]}$ via a similarity transformation, and likewise C_d is obtained from $\tilde{C}_d$ via a similarity transformation. In particular, B_0 is the scalar 1 because $A_{[0]}$ is the scalar 1. Finally, since $\mathcal{B}$ is an *orthonormal* basis, we have that $[T_d]_{\mathcal{B}} = ([T^*_d]_{\mathcal{B}})^*$, so $[T_d]_{\mathcal{B}}$ has the form given in (3.22). It therefore only remains to note that $C_d = [T_d|_{E_{\kappa-1}}]_{\mathcal{B}_E}$ simply because $\mathcal{B} = \{\tilde{e}^*_\alpha : 0 \le |\alpha| < \kappa\} \cup \mathcal{B}_E$ and because $E_{\kappa-1}$ is right-invariant under T_d.

(d) Because the spaces E_s are nested,

$$\mathcal{B}_s = \{\tilde{e}^*_\alpha : s < |\alpha| < \kappa\} \cup \mathcal{B}_E$$

is an orthonormal basis for E_s. It therefore follows from (3.22) that $[T_d|_{E_s}]_{\mathcal{B}_s}$ is that bottom right submatrix of the right-side of (3.22) which has the blocks $B_{s+1}, \dots, B_{\kappa-1}, C_d$ on the diagonal. In particular, the operators $T_d|_{E_0}$ are simultaneously block lower-triangularized in the basis $\mathcal{B}_0$, with the blocks $B_1, \dots, B_{\kappa-1}, C_d$ appearing along the diagonal of $[T_d|_{E_0}]_{\mathcal{B}_0}$. It therefore follows easily from (2.26) that

$$\hat{\rho}_\infty(\{T_d|_{E_0}\}_{d\in D}) = \max\{\rho(B_1), \dots, \rho(B_{\kappa-1}), \hat{\rho}_\infty(\{C_d\}_{d\in D})\}.$$

However, the eigenvalues of B_s and $A_{[s]}^{-1}$ coincide, and by [**CHM98**, Lemma 4.2], the spectrum of $A_{[s]}^{-1}$ is $\{\lambda^{-\alpha} : |\alpha| = s\}$, where $\lambda = (\lambda_1, \dots, \lambda_n)^{\mathrm{T}}$ is the vector of all eigenvalues of A. Since $|\lambda_i| > 1$ for each i, we therefore have

$$\begin{aligned} \max\{\rho(B_1), \dots, \rho(B_{\kappa-1})\} &= \max\{|\lambda^{-\alpha}| : 1 \le |\alpha| \le \kappa - 1\} \\ &= \max\{|\lambda_1^{-1}|, \dots, |\lambda_n^{-1}|\} \\ &= \rho(A^{-1}). \end{aligned}$$

The result then follows since $C_d = [T_d|_{E_{\kappa-1}}]_{\mathcal{B}_E}$. □

Note from (3.25) that for the digit $d = 0$, we have

$$[e_\alpha T_d]_{|\alpha|=s} = A_{[s]}^{-1} [e_\alpha]_{|\alpha|=s},$$

since $Q_{[s,t]}(0) = 0$ when $t < s$. If A is diagonalizable, then it is possible to make a change of basis so that $A_{[s]}^{-1} = \operatorname{diag}(\lambda^{-\alpha} : |\alpha| = s)$. Hence, in this basis the vectors e_α are left $\lambda^{-\alpha}$-eigenvectors of T_0. However, even in this basis they are not eigenvectors of those T_d with $d \neq 0$.

3.6. Necessary Conditions for the Existence of Continuous Vector Scaling Functions

We saw in Theorem 3.4 that if the coefficients c_k of the refinement equation satisfy the conditions for minimal accuracy, then a sufficient condition for the existence of a continuous vector scaling function is that $\hat{\rho}(\{T_d|_{E_0}\}_{d\in D}) < 1$.

The matrices $T_d = [c_{Ai-j+d}]_{i,j\in\Omega}$ and the subspace $E_0 = (e_0^*)^\perp$ depend implicitly on the choice of $\Omega \subset \Gamma$. In this section we will show that if the minimal accuracy conditions are satisfied and if in addition the lattice translates of f are "stable" and the set Ω is "admissible," then the condition $\hat{\rho}(\{T_d|_{E_0}\}_{d\in D}) < 1$ is also necessary for the existence of a continuous vector scaling function.

The definition of "stable translates" that we shall use is as follows.

Definition 3.18. A vector function $g \in L^\infty(\mathbf{R}^n, \mathbf{C}^r)$ is said to have *L^∞-stable translates* if there exist constants C_1, $C_2 > 0$ such that

$$C_1 \sup_{k\in\Gamma} \max_i |a_{k,i}| \le \Big\| \sum_{k\in\Gamma} a_k\, g(x+k) \Big\|_{L^\infty} \le C_2 \sup_{k\in\Gamma} \max_i |a_{k,i}|$$

for all finitely supported sequences of row vectors $\{a_k\}_{k\in\Gamma}$, where we write $a_k = (a_{k,1}, \dots, a_{k,r}) \in C^{1\times r}$ for $k \in \Gamma$. ◇

Using the fact that all norms on a finite-dimensional vector space are equivalent, it is easy to see that the definition of L^∞-stability can be recast into the following form.

Lemma 3.19. *Let $\|\cdot\|$ be any norm on $\mathbf{C}^{r\times r}$. Then a vector function $g \in L^\infty(\mathbf{R}^n, \mathbf{C}^r)$ has L^∞-stable translates if and only if there exist constants C_1, $C_2 > 0$ such that*

$$C_1 \sup_{k\in\Gamma} \|B_k\| \le \Big\| \sum_{k\in\Gamma} B_k\, g(x+k) \Big\|_{L^\infty} \le C_2 \sup_{k\in\Gamma} \|B_k\|$$

for all finitely supported sequences of matrices $\{B_k\}_{k\in\Gamma}$, *where* $B_k \in C^{r\times r}$ *for* $k \in \Gamma$.

The notion of "admissible set" that we shall use is as follows.

DEFINITION 3.20. Let H be a finite subset of Γ. Then we say that a nonempty, finite set $\Omega \subset \Gamma$ is *H-admissible* if

$$w_H(\Omega) \cap \Gamma \subset \Omega,$$

where $w_H(\Omega) = A^{-1}(\Omega + H)$ is as defined in (2.6). ◇

The notion of admissibility arises naturally in the study of refinement equations. For example, if Ω is Λ-admissible then the space

$$\ell(\Omega) = \{a = [a_k]_{k\in\Gamma} \in (\mathbf{C}^{r\times 1})^{\Gamma\times 1} : \mathrm{supp}(a) \subset \Omega\}$$

is right-invariant under the infinite matrix $L = [c_{Ai-j}]_{i,j\in\Gamma}$ which appears in the statement of Theorem 3.12, and the right-eigenvectors of L corresponding to nonzero eigenvalues are necessarily elements of $\ell(\Omega)$. The eigenvalues and eigenvectors of L are intimately tied to the accuracy of the vector scaling function, a topic which is explored in more detail in [**CHM00**].

In this section, we will need to consider sets $\Omega \subset \Gamma$ which are admissible with respect to the set

$$\Lambda' = \Lambda - D = \{k - d : k \in \Lambda,\ d \in D\}$$

Since we have assumed that $0 \in D$, it follows that $\Lambda \subset \Lambda'$. The set Λ' has already made an appearance, in particular, we showed in Proposition 2.7 that the set $\Omega_{\Lambda'} = K_{\Lambda'} \cap \Gamma$ satisfies $K_\Lambda \subset Q + \Omega_{\Lambda'}$. It is easy to prove that this set $\Omega_{\Lambda'}$ is both Λ-admissible and Λ'-admissible. For clarity, we shall from now on write out the symbols $\Lambda - D$ instead of using the abbreviation Λ'.

For later use, we remark that if $\Omega \subset \Gamma$, then

$$\Omega \subset \Gamma \text{ is } (\Lambda - D)\text{-admissible} \iff A^{-1}(\Lambda + \Omega - D) \cap \Gamma \subset \Omega. \qquad (3.26)$$

Further, by [**CHM00**, Lemma 3], every finite subset of Γ is contained in a finite $(\Lambda - D)$-admissible set.

Using the above notation, we can now formulate the major result of this section as follows.

THEOREM 3.21. *Let f be a continuous, compactly supported solution to the refinement equation* (1.1) *such that f has L^∞-stable translates. Assume that the hypotheses of* Lemma 3.2(a) *are satisfied, i.e., there exists a row vector $u_0 \in C^{1\times r}$ such that*

$$u_0\hat{f}(0) \neq 0 \qquad \text{and} \qquad u_0 = \sum_{k\in\Gamma_d} u_0 c_k \ \text{for } d \in D.$$

If $\Omega \subset \Gamma$ is any $(\Lambda - D)$-admissible set such that $K_\Lambda \subset Q + \Omega$, then $\hat{\rho}(\{T_d|_{E_0}\}_{d\in D}) < 1$.

We will break the proof of Theorem 3.21 into two steps. First, we will prove that the existence of a continuous solution to the refinement equation with stable translates implies that a matrix-valued version of the cascade algorithm converges pointwise everywhere when a specific starting function is used. Second, we will prove that the convergence of this version of the cascade algorithm necessarily implies that the JSR in question is less than 1. Each of these stages is of interest in itself. Moreover, the first stage requires the assumption of stable translates but does not require any admissibility assumptions on the set Ω, while the second stage requires that the set Ω be $(\Lambda - D)$-admissible but does not require that f have stable translates.

The matrix version of the cascade algorithm referred to above is defined as follows. Let $\tilde{Q}$ be the subset of Q constructed in Proposition 2.10. This set $\tilde{Q}$ has the property that the Γ-translates of $\tilde{Q}$ cover $\mathbf{R}^n$ without overlaps. Further, $\tilde{Q}$ contains a unique element γ_0 of Γ, i.e.,

$$\tilde{Q} \cap \Gamma \;=\; \{\gamma_0\}.$$

Define

$$\varphi^{(0)}(x) \;=\; \chi_{\tilde{Q}-\gamma_0}(x) \cdot I_r, \tag{3.27}$$

where I_r is the $r \times r$ identity matrix, and let $\varphi^{(i)} \in L^\infty(\mathbf{R}^n, \mathbf{C}^{r\times r})$ be obtained by iterating the refinement operator S, i.e.,

$$\varphi^{(i+1)}(x) \;=\; S\varphi^{(i)}(x) \;=\; \sum_{k\in\Lambda} c_k\, \varphi^{(i)}(Ax - k). \tag{3.28}$$

Note that we have abused notation somewhat in (3.28), since the refinement operator S is formally defined to act on vector-valued functions, while we are here applying it to matrix-valued functions. However, the abuse is slight and the intended meaning is clear. We will perform similar abuses throughout this section without further comment.

Suppose now that the coefficients c_k of the refinement equation satisfy the conditions for minimal accuracy. Specifically, these conditions are the hypotheses of Lemma 3.2(a). In this case, there exists a row vector $u_0 \in C^{1\times r}$ such that $\sum_{k\in\Gamma} u_0 f(x+k) = 1$. We will show in the following theorem that if the translates of f are L^∞-stable, then the functions $\varphi^{(i)}$ obtained via the matrix cascade algorithm converge both uniformly (i.e., in L^∞-norm) and pointwise everywhere to the matrix-valued function $f(x)u_0$ (note that this matrix has rank one for each x). It will be important for the second stage of the proof of Theorem 3.21 that this convergence is pointwise everywhere, and not merely almost everywhere. In this first stage of the proof of Theorem 3.21 we do not require any admissibility assumptions on the set Ω.

THEOREM 3.22. *Let f be a continuous, compactly supported solution to the refinement equation* (1.1) *such that f has L^∞-stable translates. Assume that the hypotheses of* Lemma 3.2(a) *are satisfied, i.e., there exists $u_0 \in C^{1\times r}$ such that $u_0\hat{f}(0) \neq 0$ and $u_0 = \sum_{k\in\Gamma_d} u_0 c_k$ for $d \in D$. Let $\varphi^{(0)}$ be the characteristic function of the unique translate of $\tilde{Q}$ that contains the origin, and let $\varphi^{(i)}$ be the ith iteration of the cascade algorithm, cf. equations* (3.27) *and* (3.28). *Then $\varphi^{(i)}$ converges uniformly and pointwise everywhere to $f(x)u_0$ as $i \to \infty$.*

PROOF. In order to distinguish between the norm of a vector and the norm of a function, we shall in this proof use the symbol $|\cdot|_p$ to denote the ℓ^p-norm on a finite-dimensional space such as $\mathbf{R}^n$ or $\mathbf{C}^r$, and use $\|\cdot\|_{L^p}$ to denote the norm on a function space such as $L^p(\mathbf{R}^n, \mathbf{C}^r)$.

By Lemma 3.2(a), f has accuracy $\kappa \geq 1$, and, in particular,

$$\sum_{k \in \Gamma} u_0 f(x+k) = 1. \tag{3.29}$$

Equality holds everywhere in this equation since f is continuous. For each $i \geq 0$, define $g^{(i)} \in L^\infty(\mathbf{R}^n, \mathbf{C}^r)$ by

$$g^{(i)}(x) \;=\; \sum_{k \in \Gamma} f(A^{-i}k)\, u_0 f(A^i x - k). \tag{3.30}$$

We claim that $g^{(i)}$ converges both uniformly and pointwise everywhere to f.

To see this, choose any $\varepsilon > 0$. Then since f is continuous and is supported in the compact set K_Λ, it is uniformly continuous. Hence, there exists a $\delta > 0$ such that

$$|x - y|_\infty \;<\; \delta \;\Longrightarrow\; |f(x) - f(y)|_\infty \;<\; \varepsilon.$$

Let i_0 be such that $\operatorname{diam}(A^{-i_0}(K_\Lambda)) < \delta$, where we measure diameter with respect to the ℓ^∞-norm on $\mathbf{R}^n$. Choose any $i \geq i_0$, and define

$$K(x) \;=\; \{k \in \Gamma : A^i x - k \in K_\Lambda\}.$$

Note that since $K_\Lambda \subset Q + \Omega$, the cardinality of $K(x)$ is bounded by the cardinality of Ω. Further, if $k \in K(x)$, then $x - A^{-i}k \in A^{-i}(K_\Lambda)$, so $|x - A^{-i}k|_\infty < \delta$. Therefore, by using (3.29) and (3.30), we have for each $x \in \mathbf{R}^n$ that

$$\begin{aligned}
|f(x) - g^{(i)}(x)|_\infty &= \left| f(x) \sum_{k \in \Gamma} u_0 f(A^i x - k) \;-\; \sum_{k \in \Gamma} f(A^{-i}k)\, u_0 f(A^i x - k) \right|_\infty \\
&= \left| \sum_{k \in \Gamma} \bigl(f(x) - f(A^{-i}k)\bigr)\, u_0 f(A^i x - k) \right|_\infty \\
&\leq \sum_{k \in K(x)} \bigl|f(x) - f(A^{-i}k)\bigr|_\infty \, |u_0 f(A^i x - k)| \\
&\leq \sum_{k \in K(x)} \varepsilon\, |u_0|_1 \, |f(A^i x - k)|_\infty \\
&\leq \varepsilon\, |u_0|_1 \, \|f\|_{L^\infty} \, \#\Omega.
\end{aligned} \tag{3.31}$$

It follows immediately that $g^{(i)}$ converges both uniformly and pointwise everywhere to f, completing the proof of our claim.

Next, an easy induction shows that

$$f(x) \;=\; \sum_{k \in \Gamma} \varphi^{(i)}(A^{-i}k)\, f(A^i x - k)$$

and

$$\varphi^{(i)}(x) \;=\; \sum_{k \in \Gamma} \varphi^{(i)}(A^{-i}k)\, \varphi^{(0)}(A^i x - k)$$

for every $i \geq 0$. In particular, $\varphi^{(i)}$ is a "step function" that is constant on each "small tile" $A^{-i}(\tilde{Q}+k)$. Since these small tiles cover $\mathbf{R}^n$ without overlaps as k varies through Γ, and since $f(x)u_0$ is uniformly continuous, to show that $\varphi^{(i)}(x) \to f(x)u_0$ uniformly and pointwise everywhere it suffices to prove that

$$\sup_{k \in \Gamma} \left|\varphi^{(i)}(A^{-i}k) - f(A^{-i}k)u_0\right|_\infty \;\to\; 0.$$

However, we have by hypothesis that f has L^∞-stable translates, so it follows from Lemma 3.19 with $B_k = \varphi^{(i)}(A^{-i}k) - f(A^{-i}k)u_0$ that, for some constant $C > 0$,

$$\begin{aligned}
&\sup_{k\in\Gamma} \left|\varphi^{(i)}(A^{-i}k) - f(A^{-i}k)u_0\right|_\infty \\
&\qquad \leq C \left\| \sum_{k\in\Gamma} \left(\varphi^{(i)}(A^{-i}k) - f(A^{-i}k)u_0\right) f(A^i x - k) \right\|_{L^\infty} \\
&\qquad = C\, \|f - g^{(i)}\|_{L^\infty} \\
&\qquad \to 0.
\end{aligned}$$

□

Before presenting the second stage of the proof of Theorem 3.21, we require some auxiliary notation and results. We shall in the remainder of this section often encounter nested sets of the form

$$\Omega \subset \tilde{\Omega} \subset \tilde{\tilde{\Omega}} \subset \Gamma.$$

When dealing with such sets, we will use a tilde or double-tilde to denote the analogues for $\tilde{\Omega}$ or $\tilde{\tilde{\Omega}}$ of objects implicitly associated with Ω. For example, in the list following we show several objects implicitly associated with Ω and the corresponding counterparts implicitly associated with $\tilde{\Omega}$:

$$\begin{aligned}
T_d &= [c_{Ai-j+d}]_{i,j\in\Omega}, & \tilde{T}_d &= [c_{Ai-j+d}]_{i,j\in\tilde{\Omega}}, \\
e_0 &= (u_0)_{k\in\Omega} \in (\mathbf{C}^{1\times r})^{1\times\Omega}, & \tilde{e}_0 &= (u_0)_{k\in\tilde{\Omega}} \in (\mathbf{C}^{1\times r})^{1\times\tilde{\Omega}}, \\
E_0 &= (e_0^*)^\perp \subset (\mathbf{C}^{r\times 1})^{\Omega\times 1}, & \tilde{E}_0 &= (\tilde{e}_0^*)^\perp \subset (\mathbf{C}^{r\times 1})^{\tilde{\Omega}\times 1}, \\
\Phi g(x) &= [g(x+k)]_{k\in\Omega}, & \tilde{\Phi} g(x) &= [g(x+k)]_{k\in\tilde{\Omega}},
\end{aligned}$$

and so forth. The need for these larger sets $\tilde{\Omega}$ and $\tilde{\tilde{\Omega}}$ arises because we will be applying the cascade algorithms to functions that are compactly supported but which need not be supported within the attractor K_Λ. The next lemma allows us to control the supports of the iterates of the cascade algorithm by observing that these supports must converge in Hausdorff metric to K_Λ. For this purpose, recall the notation introduced in Section 2.2 in association with the Hausdorff metric, specifically the definition from (2.4) that

$$B_\varepsilon = \{x \in \mathbf{R}^n : \operatorname{dist}(x, B) < \varepsilon\}.$$

LEMMA 3.23. *Let $\Omega \subset \tilde{\Omega} \subset \Gamma$ be such that*

(i) $K_\Lambda \subset Q + \Omega$,

(ii) $\tilde{\Omega}$ *is $(\Lambda - D)$-admissible.*

If g is any function such that $\operatorname{supp}(g) \subset Q + \tilde{\Omega}$, *then* $\operatorname{supp}(Sg) \subset Q + \tilde{\Omega}$ *as well. Further, given $\varepsilon > 0$ there exists $i_0 > 0$ such that* $\operatorname{supp}(S^i g) \subset (Q + \Omega)_\varepsilon$ *for all $i \geq i_0$.*

PROOF. Suppose that $Sg(x) \neq 0$. Since $Sg(x) = \sum_{k\in\Lambda} c_k\, g(Ax-k)$, there must exist some $k \in \Lambda$ such that $Ax - k \in Q + \tilde{\Omega}$. Hence $Ax = y + k + \ell$ for some $y \in Q$ and $\ell \in \tilde{\Omega}$. The point $k + \ell$ must lie in some coset $\Gamma_d = A(\Gamma) - d$, so $k + \ell = Aj + d$ for some $j \in \Gamma$ and $d \in D$. Since $\tilde{\Omega}$ is $(\Lambda - D)$-admissible, we therefore have

$$j \;=\; A^{-1}(k+\ell-d) \;\in\; A^{-1}(\Lambda+\tilde{\Omega}-D) \cap \Gamma \;\subset\; \tilde{\Omega}.$$

Note that $A^{-1}(y+d) \in Q$ since $y \in Q$ and $d \in D$ and Q is the attractor $Q = K_D = A^{-1}(K_D + D)$. Therefore,

$$x \;=\; A^{-1}(y+k+\ell) \;=\; A^{-1}(y+d) \;+\; A^{-1}(k+\ell-d) \;\in\; Q+\tilde{\Omega}.$$

Since $Q + \tilde{\Omega}$ is compact, we conclude that $\operatorname{supp}(Sg) \subset Q + \tilde{\Omega}$.

Now let $\varepsilon > 0$ be given. Since K_Λ is the attractor of the IFS generated by $\{w_k\}_{k\in\Lambda}$, for any nonempty compact set $B \subset \mathbf{R}^n$ the sequence of sets $w_\Lambda^i(B)$ must converge to K_Λ in the Hausdorff metric as $i \to \infty$. In particular, for all i large enough we must have

$$\operatorname{supp}(S^i g) \;\subset\; w_\Lambda^i(\operatorname{supp}(g)) \;\subset\; (K_\Lambda)_\varepsilon \;\subset\; (Q+\Omega)_\varepsilon. \quad \square$$

Next, we observe that by choosing a convenient ordering of $\tilde{\Omega}$, we can place the large matrix $\tilde{T}_d$ into a block diagonal form in which the smaller matrix T_d appears on the diagonal.

LEMMA 3.24. *Let $\Omega \subset \tilde{\Omega} \subset \Gamma$ be such that Ω is $(\Lambda - D)$-admissible. Let $\tilde{\Omega}$ be ordered so that the elements of Ω precede the elements of $\tilde{\Omega} \setminus \Omega$. Then there exist matrices B_d, C_d such that $\tilde{T}_d = [c_{Ai-j+d}]_{i,j\in\tilde{\Omega}}$ has the block form*

$$\tilde{T}_d \;=\; \begin{bmatrix} T_d & B_d \\ 0 & C_d \end{bmatrix}.$$

PROOF. Since $T_d = [c_{Ai-j+d}]_{i,j\in\Omega}$ and we have chosen an ordering of $\tilde{\Omega}$ in which the elements of Ω are listed first, we can certainly write $\tilde{T}_d$ in the block form

$$\tilde{T}_d \;=\; \begin{bmatrix} T_d & B_d \\ A_d & C_d \end{bmatrix}.$$

Thus, our goal is show that $A_d = [c_{Ai-j+d}]_{i\in\tilde{\Omega}\setminus\Omega,\, j\in\Omega}$ is the zero matrix. Therefore, let $j \in \Omega$ be fixed, and suppose that $c_{Ai-j+d} \neq 0$ for some $i \in \Gamma$. Then we must have $Ai - j + d \in \Lambda$. Since Ω is $(\Lambda - D)$-admissible, it therefore follows that

$$i \;\in\; A^{-1}(\Lambda+j-d) \cap \Gamma \;\subset\; A^{-1}(\Lambda+\Omega-D) \cap \Gamma \;\subset\; \Omega,$$

which proves that $A_d = 0$. $\square$

The following result is similar in nature to Proposition 2.13. The restriction in Proposition 2.13 that the support of g be contained in the attractor K_Λ is here relaxed to requiring only that $\operatorname{supp}(g)$ be contained in some possibly larger set $Q+\tilde{\Omega}$. The cost is that additional restrictions must be placed on $\tilde{\Omega}$, and furthermore, the conclusion holds only for large enough iterations instead of for all iterations.

PROPOSITION 3.25. *Let $\Omega \subset \tilde{\Omega} \subset \Gamma$ be such that*

(i) $K_\Lambda \subset Q + \Omega$,

(ii) Ω *and* $\tilde{\Omega}$ *are both* $(\Lambda - D)$*-admissible, and*

(iii) $(Q+\Omega)_\varepsilon \subset Q + \tilde{\Omega}$.

Let $g\colon \mathbf{R}^n \to \mathbf{C}^r$ *be any function such that* $\operatorname{supp}(g) \subset Q + \tilde{\Omega}$. *Then there exists* $i_0 > 0$ *such that if* $x \in Q$ *and* $x = .\varepsilon_1\varepsilon_2\cdots$ *is any A-nary expansion of* x, *then*

$$\forall\, i \geq i_0, \quad \tilde{\Phi} S^i g(x) \;=\; \tilde{T}_{\varepsilon_1} \cdots \tilde{T}_{\varepsilon_i} \tilde{\Phi} g(y_i), \tag{3.32}$$

where

$$y_i \;=\; .\varepsilon_{i+1}\varepsilon_{i+2}\cdots \in Q.$$

Consequently,

$$\forall\, i \geq i_0, \quad \tilde{\Phi} S^i g \;=\; \tilde{T}^i \tilde{\Phi} g.$$

PROOF. Let $\tilde{\tilde{\Omega}}$ be finite but large enough that we have both

$$\tilde{\Omega} \subset \tilde{\tilde{\Omega}} \subset \Gamma \qquad \text{and} \qquad Q + \tilde{\Omega} \subset (Q + \tilde{\tilde{\Omega}})^\circ.$$

Suppose that $g\colon \mathbf{R}^n \to \mathbf{C}^r$ satisfies $\operatorname{supp}(g) \subset Q + \tilde{\Omega}$. Then by Lemma 3.23 there is an $i_0 > 0$ such that $\operatorname{supp}(S^i g) \subset (Q+\Omega)_\varepsilon$ for all $i \geq i_0$.

Now choose $x \in Q$, and let $x = .\varepsilon_1\varepsilon_2\cdots$ be any particular A-nary expansion of x. Then $y_1 = .\varepsilon_2\varepsilon_3\cdots = Ax - \varepsilon_1 \in Q$. If $g(y_1 + k) \neq 0$ for some $k \in \Gamma$, then we must have $y_1 + k \in \operatorname{supp}(g) \subset Q + \tilde{\Omega} \subset (Q + \tilde{\tilde{\Omega}})^\circ$. Lemma 2.9, applied to the set $\tilde{\tilde{\Omega}}$ instead of Ω, therefore implies that $k \in \tilde{\tilde{\Omega}}$. A calculation identical to the one in (2.23), except with Ω replaced by $\tilde{\tilde{\Omega}}$, shows that

$$\tilde{\tilde{\Phi}} S g(x) \;=\; \tilde{\tilde{T}}_{\varepsilon_1} \tilde{\tilde{\Phi}} g(y_1). \tag{3.33}$$

By Lemma 3.23, we have that $\operatorname{supp}(Sg) \subset Q + \tilde{\Omega}$, so we can iterate the calculation in (3.33) to obtain

$$\tilde{\tilde{\Phi}} S^i g(x) \;=\; \tilde{\tilde{T}}_{\varepsilon_1} \cdots \tilde{\tilde{T}}_{\varepsilon_i} \tilde{\tilde{\Phi}} g(y_i). \tag{3.34}$$

Choose now any ordering of $\tilde{\tilde{\Omega}}$ such that the elements of $\tilde{\Omega}$ precede the elements of $\tilde{\tilde{\Omega}} \setminus \tilde{\Omega}$. Then Lemma 3.24, applied to the sets $\tilde{\Omega} \subset \tilde{\tilde{\Omega}}$ instead of $\Omega \subset \tilde{\Omega}$, implies that $\tilde{\tilde{T}}_d$ has the block form

$$\tilde{\tilde{T}}_d \;=\; \begin{bmatrix} \tilde{T}_d & \tilde{B}_d \\ 0 & \tilde{C}_d \end{bmatrix} \tag{3.35}$$

for some matrices $\tilde{B}_d$ and $\tilde{C}_d$. We claim that the folding $\tilde{\tilde{\Phi}} S^i g(y) = [S^i g(y+k)]_{k \in \tilde{\tilde{\Omega}}}$ similarly has the block form

$$\tilde{\tilde{\Phi}} S^i g(y) \;=\; \begin{bmatrix} \tilde{\Phi} S^i g(y) \\ 0 \end{bmatrix}, \qquad \text{for } y \in Q \text{ and } i \geq i_0. \tag{3.36}$$

To show this we simply have to show that if $y \in Q$ and $i \geq i_0$, then $S^i g(y+k) = 0$ for $k \in \tilde{\tilde{\Omega}} \setminus \tilde{\Omega}$. However, if $i \geq i_0$, then $\operatorname{supp}(S^i g) \subset (Q+\Omega)_\varepsilon \subset (Q + \tilde{\Omega})^\circ$. Lemma 2.9, applied to the set $\tilde{\Omega}$ instead of Ω, therefore implies that if $S^i g(y+k) \neq 0$ then $k \in \tilde{\Omega}$. Hence (3.36) is valid.

Finally, combining (3.34), (3.35), and (3.36), we see that for $i \geq i_0$,

$$\begin{aligned}
\begin{bmatrix} \tilde{\Phi} S^i g(x) \\ 0 \end{bmatrix} &= \tilde{\tilde{\Phi}} S^i g(x) \\
&= \tilde{\tilde{T}}_{\varepsilon_1} \cdots \tilde{\tilde{T}}_{\varepsilon_i} \tilde{\tilde{\Phi}} g(y_i) \\
&= \begin{bmatrix} \tilde{T}_{\varepsilon_1} \cdots \tilde{T}_{\varepsilon_i} & * \\ 0 & * \end{bmatrix} \begin{bmatrix} \tilde{\Phi} S^i g(y_i) \\ 0 \end{bmatrix} \\
&= \begin{bmatrix} \tilde{T}_{\varepsilon_1} \cdots \tilde{T}_{\varepsilon_i} \tilde{\Phi} S^i g(y_i) \\ 0 \end{bmatrix},
\end{aligned}$$

from which the result follows. □

Now we can complete the second stage of the proof of Theorem 3.21. Specifically, we show next that the pointwise convergence of the matrix cascade algorithm implies a restriction on the uniform JSR. Note that this result does not require that f have L^∞-stable translates.

THEOREM 3.26. *Let f be a continuous, compactly supported solution to the refinement equation* (1.1). *Assume that the hypotheses of* Lemma 3.2(a) *are satisfied, i.e., there exists $u_0 \in \mathbf{C}^{1\times r}$ such that $u_0 \hat{f}(0) \neq 0$ and $u_0 = \sum_{k\in\Gamma_d} u_0 c_k$ for $d \in D$. Let Ω be any $(\Lambda - D)$-admissible subset of Γ such that $K_\Lambda \subset Q + \Omega$. If the functions $\varphi^{(i)}$ defined by* (3.27) *and* (3.28) *converge pointwise everywhere to $f(x)u_0$, then $\hat{\rho}(\{T_d|_{E_0}\}_{d\in D}) < 1$.*

PROOF. By [**CHM00**, Lemma 4.7], every finite subset of Γ is contained in an admissible set. Hence, if we fix an $\varepsilon > 0$, then we can find a $(\Lambda - D)$-admissible set $\tilde{\Omega}$ such that

$$\Omega \subset \tilde{\Omega} \subset \Gamma \qquad \text{and} \qquad (Q+\Omega)_\varepsilon \subset (Q+\tilde{\Omega})^\circ.$$

We will prove that if $\{\varepsilon_i\}_{i=1}^\infty$ is any sequence of digits $\varepsilon_i \in D$, then the matrix product $T_{\varepsilon_1} \cdots T_{\varepsilon_i}$ converges as $i \to \infty$ to the rank-one matrix each of whose columns is $\Phi(f(x)u_0)$, where x is the point

$$x = .\varepsilon_1\varepsilon_2\cdots = \sum_{j=1}^\infty A^{-j}\varepsilon_j \in Q.$$

This will occupy us for the majority of the proof of the theorem. From this fact we will then deduce that $\hat{\rho}(\{T_d|_{E_0}\}_{d\in D}) < 1$.

To begin, let a sequence of digits $\{\varepsilon_i\}_{i=1}^\infty$ be fixed, and set $x = .\varepsilon_1\varepsilon_2\cdots \in Q$. By hypothesis, $\varphi^{(i)}(x) \to f(x)u_0$ when $\varphi^{(0)}(x) = \chi_{\tilde{Q}-\gamma_0}(x) \cdot I_r$. Let σ_h denote the translation operator, i.e., $\sigma_h g(x) = g(x-h)$. For each $h \in \tilde{\Omega}$, set

$$\varphi_h^{(0)}(x) = \big(\sigma_{h+\gamma_0}\varphi^{(0)}\big)(x) = \chi_{\tilde{Q}+h}(x) \cdot I_r, \tag{3.37}$$

and define

$$\varphi_h^{(i)}(x) = S^i \varphi_h^{(0)}(x) = S^i\big(\sigma_{h+\gamma_0}\varphi^{(0)}\big)(x) = \sigma_{A^{-i}(h+\gamma_0)}(S^i\varphi^{(0)})(x). \tag{3.38}$$

Note that $\text{supp}(\varphi_h^{(0)}) \subset Q + h \subset Q + \tilde{\Omega}$ for each $h \in \tilde{\Omega}$. Lemma 3.23 therefore implies that $\text{supp}(\varphi_h^{(i)}) \subset Q + \tilde{\Omega}$ for all i, and moreover that $\text{supp}(\varphi_h^{(i)}) \subset (Q+\Omega)_\varepsilon$ for all i large enough. Since this is true for each h in the finite set $\tilde{\Omega}$, there is some i_0 such that

$$\forall\, h \in \tilde{\Omega}, \quad \forall\, i \geq i_0, \quad \text{supp}(\varphi_h^{(i)}) \;\subset\; (Q+\Omega)_\varepsilon.$$

Now fix any particular $h \in \Omega$ (*not* merely $h \in \tilde{\Omega}$ but specifically $h \in \Omega$). Consider the points

$$y_i \;=\; .\varepsilon_{i+1}\varepsilon_{i+2}\cdots \;\in\; Q.$$

Recall that $\tilde{Q}$ was defined to have the property that the Γ-translates of $\tilde{Q}$ cover $\mathbf{R}^n$ without overlaps. Therefore, the point $y_i + h$ must lie in some unique translate of $\tilde{Q}$. Hence, there exist unique points $q_i \in \tilde{Q}$ and $k_i \in \Gamma$ such that

$$y_i + h \;=\; q_i + k_i. \tag{3.39}$$

Note that

$$q_i + k_i \;=\; y_i + h \;\in\; Q + \Omega \;\subset\; (Q+\tilde{\Omega})^\circ.$$

Lemma 2.9, applied to the set $\tilde{\Omega}$ instead of Ω, therefore implies that $k_i \in \tilde{\Omega}$. Hence, if we let $\delta_{h,j}$ denote the Kronecker delta, then then folding of $\varphi_{k_i}^{(0)}$ satisfies

$$\begin{aligned}
\tilde{\Phi}\varphi_{k_i}^{(0)}(y_i) &= \Big[\varphi_{k_i}^{(0)}(y_i+j)\Big]_{j\in\tilde{\Omega}} && \text{by definition of } \tilde{\Phi}\\
&= \Big[\chi_{\tilde{Q}}(y_i+j-k_i)\cdot I_r\Big]_{j\in\tilde{\Omega}} && \text{by (3.37)}\\
&= \Big[\chi_{\tilde{Q}}(q_i-h+j)\cdot I_r\Big]_{j\in\tilde{\Omega}} && \text{by (3.39)}\\
&= \big[\delta_{h,j}\cdot I_r\big]_{j\in\tilde{\Omega}} && \text{by Proposition 2.10.}
\end{aligned}$$

Fix any ordering on $\tilde{\Omega}$ such that the elements of Ω precede the elements of $\tilde{\Omega}\setminus\Omega$. Define

$$\Delta_h \;=\; \big[\delta_{h,j}\cdot I_r\big]_{j\in\Omega} \qquad \text{and} \qquad \tilde{\Delta}_h \;=\; \big[\delta_{h,j}\cdot I_r\big]_{j\in\tilde{\Omega}} \;=\; \begin{bmatrix}\Delta_h\\ 0\end{bmatrix}.$$

That is, Δ_h and $\tilde{\Delta}_h$ are column vectors with the identity block I_r appearing in "row block h" and zeros elsewhere. Multiplication of a matrix by Δ_h or $\tilde{\Delta}_h$ on the right therefore selects out "column block h" from that matrix. Thus, by Proposition 3.25, and equation (3.32) in particular, we have for $i \geq i_0$ that

$$\tilde{\Phi}\varphi_{k_i}^{(i)}(x) \;=\; \tilde{\Phi}S^i\varphi_{k_i}^{(0)}(x) \;=\; \tilde{T}_{\varepsilon_1}\cdots\tilde{T}_{\varepsilon_i}\tilde{\Phi}\varphi_{k_i}^{(0)}(y_i) \;=\; \tilde{T}_{\varepsilon_1}\cdots\tilde{T}_{\varepsilon_i}\tilde{\Delta}_h \tag{3.40}$$

is "column block h" of $\tilde{T}_{\varepsilon_1}\cdots\tilde{T}_{\varepsilon_i}$. On the other hand, we have by hypothesis that

$$S^i\varphi^{(0)}(x) \;=\; \varphi^{(i)}(x) \;\to\; f(x)u_0. \tag{3.41}$$

Therefore,

$$\begin{aligned}
\tilde{T}_{\varepsilon_1}\cdots\tilde{T}_{\varepsilon_i}\tilde{\Delta}_h &= \tilde{\Phi}\varphi_{k_i}^{(i)}(x) && \text{by (3.40)}\\
&= \tilde{\Phi}(\sigma_{A^{-i}(k_i+\gamma_0)}S^i\varphi^{(0)})(x) && \text{by (3.38)}\\
&\to \tilde{\Phi}(f(x)u_0),
\end{aligned}$$

the conclusion on the preceding line following from (3.41), the contractivity of A^{-1}, and the fact that each k_i lies in the finite set $\tilde{\Omega}$. Thus, "column block h" of $\tilde{T}_{\varepsilon_1} \cdots \tilde{T}_{\varepsilon_i}$ converges to $\tilde{\Phi}(f(x)u_0)$. This is true for each $h \in \Omega$. However, the column blocks of $\tilde{T}_{\varepsilon_1} \cdots \tilde{T}_{\varepsilon_i}$ are indexed by the larger set $\tilde{\Omega}$, so let us examine the column blocks corresponding to indices in Ω in more detail. Since we have ordered $\tilde{\Omega}$ so that the elements of Ω come first, Lemma 3.24 implies that $\tilde{T}_d$ has the block form

$$\tilde{T}_d = \begin{bmatrix} T_d & B_d \\ 0 & C_d \end{bmatrix}$$

for some matrices B_d, C_d. Consequently,

$$\tilde{T}_{\varepsilon_1} \cdots \tilde{T}_{\varepsilon_i} \tilde{\Delta}_h = \begin{bmatrix} T_{\varepsilon_1} \cdots T_{\varepsilon_i} & * \\ 0 & * \end{bmatrix} \begin{bmatrix} \Delta_h \\ 0 \end{bmatrix} = \begin{bmatrix} T_{\varepsilon_1} \cdots T_{\varepsilon_i} \Delta_h \\ 0 \end{bmatrix}.$$

Further,

$$\tilde{\Phi}(f(x)u_0) = \begin{bmatrix} \Phi(f(x)u_0) \\ * \end{bmatrix},$$

so we conclude that

$$T_{\varepsilon_1} \cdots T_{\varepsilon_i} \Delta_h \to \Phi(f(x)u_0). \tag{3.42}$$

Since the columns blocks of $T_{\varepsilon_1} \cdots T_{\varepsilon_i}$ are indexed by Ω, equation (3.42) implies that each column block of $T_{\varepsilon_1} \cdots T_{\varepsilon_i}$ converges to $\Phi(f(x)u_0)$. Therefore, the product $T_{\varepsilon_1} \cdots T_{\varepsilon_i}$ converges to to the matrix $B(x)$ consisting of Ω column blocks each equal to $\Phi(f(x)u_0)$. That is,

$$T_{\varepsilon_1} \cdots T_{\varepsilon_i} \to B(x) = \big(\Phi(f(x)u_0)\big)_{k \in \Omega}.$$

This matrix $B(x)$ is rank-one because each column block $\Phi(f(x)u_0)$ consists of rows that are multiples of the $1 \times r$ row vector u_0.

Thus, we have demonstrated that $T_{\varepsilon_1} \cdots T_{\varepsilon_i}$ converges to a rank-one matrix for each sequence of digits $\{\varepsilon_i\}_{i=1}^{\infty}$. We will now show that this implies that $(T_{\varepsilon_1} \cdots T_{\varepsilon_i})|_{E_0}$ converges to the zero matrix for each such sequence of digits. The key ingredient is the hypothesis that the coefficients c_k satisfy the conditions for minimal accuracy. Because of this, Theorem 3.17 implies that there exists an orthonormal basis $\mathcal{B}$ for $(\mathbf{C}^{r \times 1})^{\Omega \times 1}$ such that each matrix has in this basis the block form

$$[T_d]_{\mathcal{B}} = \begin{bmatrix} 1 & 0 \\ * & C_d \end{bmatrix},$$

where 1 is the *scalar* 1, and $C_d = [T_d|_{E_0}]_{\mathcal{B}_0}$ is the matrix for T_d restricted to E_0 with respect to an orthonormal basis $\mathcal{B}_0$ for E_0. Consequently, working in this basis, we have for each i that

$$[T_{\varepsilon_1} \cdots T_{\varepsilon_i}]_{\mathcal{B}} = \begin{bmatrix} 1 & 0 \\ * & C_{\varepsilon_1} \cdots C_{\varepsilon_i} \end{bmatrix}.$$

Since $T_{\varepsilon_1} \cdots T_{\varepsilon_i}$ converges to a rank-one matrix, the product $C_{\varepsilon_1} \cdots C_{\varepsilon_i}$ must therefore converge to the zero matrix as $i \to \infty$. This implies by [**BW92**, Thm. I] that $\hat{\rho}(\{C_d\}_{d \in D}) < 1$, and completes the proof. □

Finally, the proof of Theorem 3.21 follows by combining Theorems 3.22 and 3.26.

3.7. Hölder Continuity

Once a vector scaling function is known to be continuous, the joint spectral radius can be used to compute the global Hölder exponent of continuity of f.

Let $|\cdot|$ be any norm on $\mathbf{R}^n$ and let $\|\cdot\|$ be any norm on $\mathbf{C}^r$. A continuous function $g\colon \mathbf{R}^n \to \mathbf{C}^r$ is *Hölder continuous* with exponent $\alpha > 0$ if there exists a constant K such that $\|g(x) - g(y)\| \leq K\,|x - y|^\alpha$ for every x and y. The value of α is independent of the choice of norms. This definition is global in the sense that the "worst" point x and the "least smooth" component g_i of g will determine the global Hölder exponent of g.

Suppose that f is a continuous, compactly supported solution to the refinement equation (1.1). Then, by Proposition 2.13,

$$\forall\, d \in D, \quad \forall\, x \in w_d(Q), \quad \Phi f(x) \;=\; T_d \Phi f(Ax - d). \tag{3.43}$$

As a consequence, the subspace

$$W_0 \;=\; \operatorname{span}\{\Phi f(x) - \Phi f(y) : x, y \in Q\}$$

is right-invariant under T_d for each $d \in D$. Note that if f satisfies the hypotheses of Lemma 3.2(a), so f has accuracy $\kappa \geq 1$ and $W_0 \subset E_0$ with E_0 defined by (3.9). In Theorem 3.4 we saw that the condition $\hat\rho_\infty(\{T_d|_{E_0}\}_{d\in D}) < 1$, with appropriate additional hypotheses, is a sufficient condition for the existence of a continuous vector scaling function f. The following result shows that the condition $\hat\rho_\infty(\{T_d|_{W_0}\}_{d\in D}) < 1$ is a necessary condition for the existence of a continuous vector scaling function, and also shows that the value of $\hat\rho_\infty(\{T_d|_{W_0}\}_{d\in D})$ bounds the value of

$$\alpha_{\text{sup}} \;=\; \sup\{\alpha : f \text{ is Hölder continuous with exponent } \alpha\}.$$

This gives a necessary condition for the existence of a continuous vector scaling function that is complementary to the necessary conditions obtained in the preceding section. In particular, this result does not require any information on whether the cascade algorithm converges, or whether f has stable translates. On the other hand, this condition is largely of theoretical value, because the space W_0 is usually difficult to determine explicitly except in case of small numbers of coefficients in the refinement equation. On the other hand, in the one-dimensional, single-function setting with minimal accuracy, it is known that $W_0 = E_0$ if and only if f has independent translates [**Sun91**], compare also [**Hei94**]. It would be interesting to know if such a characterization can also be proved in the multidimensional setting.

The spectral radius of A^{-1} will play a role in the following result. Note that A^{-1} is contractive since A is expansive, and therefore $\rho(A^{-1}) < 1$.

PROPOSITION 3.27. *Let $\Omega \subset \Gamma$ be a finite set such that $K_\Lambda \subset Q + \Omega$. If there exists a continuous, compactly supported solution $f\colon \mathbf{R}^n \to \mathbf{C}^r$ to the refinement equation* (1.1), *then $\hat\rho_\infty(\{T_d|_{W_0}\}_{d\in D}) < 1$ and*

$$\alpha_{sup} \;\leq\; \log_\sigma \hat\rho_\infty(\{T_d|_{W_0}\}_{d\in D}),$$

where $\sigma = \rho(A^{-1})$.

PROOF. Choose θ so that $\sigma < \theta < 1$. Then there exists a norm $|\cdot|$ on $\mathbf{R}^n$ such that the induced operator norm of A^{-1} satisfies $\sigma \leq |A^{-1}| < \theta < 1$. Let $\|\cdot\|$ be any norm on $(\mathbf{C}^{r\times 1})^{\Omega\times 1}$. Choose any product $\Pi = T_{\varepsilon_1}\cdots T_{\varepsilon_\ell}$, where $\varepsilon_1,\dots,\varepsilon_\ell \in D$, and let λ be any eigenvalue of $\Pi|_{W_0}$. The space

$$U_\lambda \;=\; \{w \in W_0 : (\Pi-\lambda)^k w = 0 \text{ for some } k\}$$

is right-invariant under Π. By standard Jordan techniques, there exists a subspace Z that is also right-invariant under Π and satisfies $U_\lambda \oplus Z = W_0$. Since the span of a set of vectors is the smallest subspace containing those vectors, there must exist some $x, y \in Q$ such that $\Phi f(x) - \Phi f(y) = u_\lambda + z = w$ with $0 \neq u_\lambda \in U_\lambda$ and $z \in Z$. Using Jordan arguments again, as in [**CH94**, Lemma 3], there exists a constant $C > 0$ such that

$$\|\Pi^k w\| \;\geq\; C|\lambda|^k \qquad \text{all } k > 0.$$

Let $x = .x_1x_2\cdots$ and $y = .y_1y_2\cdots$ be A-nary expansions of x and y. Define points

$$X_k = .\varepsilon_1\cdots\varepsilon_\ell\cdots\varepsilon_1\cdots\varepsilon_\ell x_1x_2\cdots \qquad \text{and} \qquad Y_k = .\varepsilon_1\cdots\varepsilon_\ell\cdots\varepsilon_1\cdots\varepsilon_\ell y_1y_2\cdots$$

in Q, where the sequence $\varepsilon_1\cdots\varepsilon_\ell$ is repeated k times. Then, using (3.43), we have

$$\|\Phi f(X_k) - \Phi f(Y_k)\| \;=\; \|\Pi^k\Phi f(x) - \Pi^k\Phi f(y)\| \;=\; \|\Pi^k w\| \;\geq\; C|\lambda|^k \tag{3.44}$$

for each $k > 0$. Since f is continuous and $|X_k - Y_k| \to 0$ as $k \to \infty$, it follows that we must have $|\lambda| < 1$. We therefore conclude from (2.26) that $\hat{\rho}_\infty(\{T_d|_{W_0}\}_{d\in D}) \leq 1$. However, (3.44) also implies that every product $(T_{\varepsilon_1}\cdots T_{\varepsilon_\ell})|_{W_0}$ must converge to zero as $\ell \to \infty$, and therefore we must in fact have $\hat{\rho}_\infty(\{T_d|_{W_0}\}_{d\in D}) < 1$.

Next we will find an upper bound for $\alpha_{\sup}$. By definition, if f has Hölder exponent α then there exists a K such that

$$\|\Phi f(X_k) - \Phi f(Y_k)\| \;\leq\; K\,|X_k - Y_k|^\alpha.$$

However,

$$|X_k - Y_k| \;=\; |A^{-\ell k}(x-y)| \;\leq\; \theta^{\ell k}\,|x-y|,$$

so

$$C\,|\lambda|^k \;\leq\; \|\Phi f(X_k) - \Phi f(Y_k)\| \;\leq\; K\,|X_k - Y_k|^\alpha \;\leq\; K\,|x-y|^\alpha\,\theta^{\alpha\ell k}.$$

Therefore, for each $k > 0$ we have

$$|\lambda|^{1/\ell} \;\leq\; \left(\frac{K\,|x-y|^\alpha}{C}\right)^{1/\ell k}\theta^\alpha.$$

Letting $k \to \infty$, we conclude that $|\lambda|^{1/\ell} \leq \theta^\alpha$. Since this is true for every eigenvalue λ of every product Π of length ℓ, it follows from (2.26) that $\hat{\rho}_\infty(\{T_d|_{W_0}\}_{d\in D}) \leq \theta^\alpha$. As this is true for every $\theta > \sigma$, we must have $\hat{\rho}_\infty(\{T_d|_{W_0}\}_{d\in D}) \leq \sigma^\alpha$, and therefore $\alpha \leq \log_\sigma \hat{\rho}_\infty(\{T_d|_{W_0}\}_{d\in D})$. □

CHAPTER 4

Multiresolution Analysis

4.1. Multiresolution Analysis

In this section we will give the definition and basic properties of multiresolution analyses of arbitrary multiplicity with respect to an arbitrary dilation matrix.

DEFINITION 4.1. A *multiresolution analysis* (MRA) *of multiplicity* r associated with a dilation matrix A is a sequence of closed subspaces $\{\mathcal{V}_j\}_{j\in\mathbf{Z}}$ of $L^2(\mathbf{R}^n)$ which satisfy:

(P1) $\mathcal{V}_j \subset \mathcal{V}_{j+1}$ for each $j \in \mathbf{Z}$,

(P2) $g(x) \in \mathcal{V}_j \iff g(Ax) \in \mathcal{V}_{j+1}$ for each $j \in \mathbf{Z}$,

(P3) $\bigcap_{j\in\mathbf{Z}} \mathcal{V}_j = \{0\}$,

(P4) $\bigcup_{j\in\mathbf{Z}} \mathcal{V}_j$ is dense in $L^2(\mathbf{R}^n)$, and

(P5) there exist functions $\varphi_1, \ldots, \varphi_r \in L^2(\mathbf{R}^n)$ such that the collection of lattice translates

$$\{\varphi_i(x-k)\}_{k\in\Gamma,\, i=1,\ldots,r} \tag{4.1}$$

forms an orthonormal basis for $\mathcal{V}_0$.

If these conditions are satisfied, then the vector function $\varphi = (\varphi_1, \ldots, \varphi_r)^{\mathrm{T}}$ is referred to as a *vector scaling function* for the MRA. ◊

The definition of multiresolution analysis can be generalized to allow the collection of lattice translates of the functions φ_i to form merely a Riesz basis instead of an orthonormal basis for $\mathcal{V}_0$. This leads then to biorthogonal wavelet bases for $L^2(\mathbf{R}^n)$. Since we are interested mostly in orthonormal wavelet bases in this manuscript, we will not consider this generalization.

The usual technique for constructing a multiresolution analysis is to start from a vector function $\varphi = (\varphi_1, \ldots, \varphi_r)^{\mathrm{T}}$ such that $\{\varphi_i(x-k)\}_{k\in\Gamma,\, i=1,\ldots,r}$ is an orthonormal system in $L^2(\mathbf{R}^n)$, and then to construct the subspaces $\mathcal{V}_j \subset L^2(\mathbf{R}^n)$ appropriately. This is made precise in the following definition. For simplicity, we shall from now on write that φ *has orthonormal lattice translates* when we mean to say that $\{\varphi_i(x-k)\}_{k\in\Gamma,\, i=1,\ldots,r}$ is an orthonormal system in $L^2(\mathbf{R}^n)$.

DEFINITION 4.2. Assume that $\varphi \in L^2(\mathbf{R}^n, \mathbf{C}^r)$ has orthonormal lattice translates. Let $\mathcal{V}_0$ be the closed linear span of the translates of the component functions φ_i, i.e.,

$$\mathcal{V}_0 \;=\; \overline{\operatorname{span}}\{\varphi_i(x-k)\}_{k\in\Gamma,\, i=1,\ldots,r}. \tag{4.2}$$

Then, for each $j \in \mathbf{Z}$, define $\mathcal{V}_j$ to be the set of all dilations of functions in $\mathcal{V}_0$ by A^j, i.e.,

$$\mathcal{V}_j \;=\; \{g(A^j x) : g \in \mathcal{V}_0\}. \tag{4.3}$$

If $\{\mathcal{V}_j\}_{j\in\mathbf{Z}}$ defined in this way forms a multiresolution analysis for $L^2(\mathbf{R}^n)$ then we say that it is the *MRA generated by* φ. ◇

EXAMPLE 4.3. In one dimension, the box function $\varphi = \chi_{[0,1)}$ generates a multiresolution analysis for $L^2(\mathbf{R})$. This MRA is usually referred to as the *Haar multiresolution analysis*, because the wavelet basis it determines is the classical Haar system $\{2^{n/2}\psi(2^n x - k)\}_{n,k\in\mathbf{Z}}$, where $\psi = \chi_{[0,1/2)} - \chi_{[1/2,1)}$.

Gröchenig and Madych [**GM92**] proved that there is a Haar-like multiresolution analysis associated to each choice of dilation matrix A and digit set D for which the attractor $Q = K_D$ is a tile (which is the standing assumption of this manuscript). In particular, they proved that if Q is a tile then the scalar-valued function χ_Q generates a multiresolution analysis of $L^2(\mathbf{R}^n)$ of multiplicity 1. By extension of the one-dimensional terminology, this MRA is called the *Haar MRA associated with A and D*. Note that the fact that $\{\chi_Q(x-k)\}_{k\in\Gamma}$ forms an orthonormal basis for $\mathcal{V}_0$ is a restatement of the assumption that the lattice translates of the tile Q have overlaps of measure zero. Further, χ_Q is refinable because Q is self-similar and because the lattice translates of Q have overlaps of measure zero. ◇

We will characterize those φ which generate multiresolution analyses in Theorem 4.4, below. To motivate this result, note that property (P2) is achieved trivially when $\mathcal{V}_j$ is defined by (4.3). Moreover, property (P5) is simply a statement that lattice translates of φ are orthonormal. We will see in the proof of Theorem 4.4 that the fact that φ has orthonormal lattice translates implies that property (P3) is also automatically satisfied. Thus, the main problem in determining whether φ generates a multiresolution analysis is the question of when properties (P1) and (P4) are satisfied. One necessary requirement for (P1) is clear. If φ does generate a multiresolution analysis, then (P1) implies that $\varphi_i \in \mathcal{V}_0 \subset \mathcal{V}_1$ for $i = 1, \dots, r$. Since (P2) and (P5) together imply that $\{m^{1/2}\,\varphi_j(Ax-k)\}_{k\in\Gamma,\, j=1,\dots,r}$ forms an orthonormal basis for $\mathcal{V}_1$, each function φ_i must therefore equal some (possibly infinite) linear combination of the functions $\varphi_j(Ax-k)$. Consequently, the vector function φ must satisfy a refinement equation of the form

$$\varphi(x) \;=\; \sum_{k\in\Gamma} c_k\, \varphi(Ax-k) \tag{4.4}$$

for some choice of $r \times r$ matrices c_k. We will only consider the case where the functions φ_i have compact support; since φ has orthonormal lattice translates, this implies that only finitely many of the matrices c_k in (4.4) can be nonzero. Hence, in this case the refinement equation in (4.4) has the same form as the refinement equation (1.1) that was studied in the preceding chapters.

THEOREM 4.4. *Assume that $\varphi = (\varphi_1, \dots, \varphi_r)^T \in L^2(\mathbf{R}^n, \mathbf{C}^r)$ is compactly supported and has orthonormal lattice translates, i.e.,*

$$\langle \varphi_i(x-k),\, \varphi_j(x-\ell)\rangle \;=\; \int \varphi_i(x-k)\, \overline{\varphi_j(x-\ell)}\, dx \;=\; \delta_{i,j}\, \delta_{k,\ell}.$$

Let $\mathcal{V}_j \subset L^2(\mathbf{R}^n)$ for $j \in \mathbf{Z}$ be defined by (4.2) and (4.3). Then the following statements hold.

(a) *Properties* (P2), (P3), *and* (P5) *are satisfied.*

(b) *Property* (P1) *is satisfied if and only if* φ *satisfies a refinement equation of the form*

$$\varphi(x) = \sum_{k\in\Lambda} c_k\, \varphi(Ax-k) \tag{4.5}$$

for some $r \times r$ *matrices* c_k *and some finite set* $\Lambda \subset \Gamma$.

(c) *If*

$$\|\hat{\varphi}(0)\|^2 = \sum_{i=1}^{r} |\hat{\varphi}_i(0)|^2 = \sum_{i=1}^{r} \left| \int \varphi_i(x)\, dx \right|^2 = |Q|, \tag{4.6}$$

then Property (P4) *is satisfied. If* φ *is refinable, i.e., if* (4.5) *holds, then Property* (P4) *is satisfied if and only if* (4.6) *holds.*

Note that the assumption that φ_i is square-integrable and compactly supported implies that $\varphi_i \in L^1(\mathbf{R}^n)$, so $\hat{\varphi}_i(0) = \int \varphi_i(x)\, dx$ is well-defined. Also recall that $|Q| = |P|$, where P is the fundamental domain for the lattice Γ defined in (2.3) (in particular, P is a rectangular parallelepiped). For example, if $\Gamma = \mathbf{Z}^n$ then we can take $P = [0,1)^n$, and therefore $|Q| = |P| = 1$.

Theorem 4.4 generalizes a result of Cohen [**Coh90**], which applied specifically to the case of multiplicity 1 and dilation $A = 2I$. Cohen's estimates used a decomposition of $\mathbf{R}^n$ into dyadic cubes, making essential use of the fact that the uniform dilation $A = 2I$ maps dyadic cubes into dyadic cubes. However, this need not be true for an arbitrary dilation matrix A, so this particular decomposition is no longer feasible. Instead, we will use a decomposition based on the tile Q, and make use of the associated Haar multiresolution analysis discussed in Example 4.3. Before we can implement this decomposition for the proof of Theorem 4.4, we require some auxiliary notation and results.

In order to deal more concisely with the dilations translations of a given function we introduce the following notation. Given a function $g\colon \mathbf{R}^n \to \mathbf{C}^r$ and given $j \in \mathbf{Z}$ and $k \in \Gamma$, we write

$$g^{j,k}(x) = m^{j/2}\, g(A^j x - k) = m^{j/2}\, g(A^j(x - A^{-j}k))$$

to denote a translation of g by $A^{-j}k$ followed by an L^2-normalized dilation of g by A^j.

Our first observation is an immediate consequence of Gröchenig and Madych's generalization of the Haar multiresolution analysis.

LEMMA 4.5. *The collection*

$$\{\chi_Q^{j,k}\}_{j\in\mathbf{Z},k\in\Gamma} = \{m^{j/2}\,\chi_Q(A^j x - k)\}_{j\in\mathbf{Z},k\in\Gamma}$$

is complete in $L^2(\mathbf{R}^n)$, *i.e., its finite linear span is dense in* $L^2(\mathbf{R}^n)$.

PROOF. Let $\{\mathcal{V}_j\}_{j\in\mathbf{Z}}$ be the Haar multiresolution analysis generated by χ_Q, as discussed in Example 4.3. Then for each fixed j, the collection of translates $\{\chi_Q^{j,k}\}_{k\in\Gamma}$ forms an orthonormal basis for the subspace $\mathcal{V}_j$. Since the union of the $\mathcal{V}_j$ is dense in $L^2(\mathbf{R}^n)$, the union of these orthonormal systems must form a complete set in $L^2(\mathbf{R}^n)$. □

Next, we will estimate the number of lattice translates of Q which lie in the interior of a dilated tile A^jQ, $j \geq 1$. Note that the fact that Q is self-similar combined with the fact that translates of Q tile $\mathbf{R}^n$ with overlaps with measure zero implies that A^jQ is a union of exactly m^j translates of Q, with each such translate lying entirely inside A^jQ (but not necessarily in the *interior* of A^jQ). Lemma 4.7 below will show that the ratio of the number of those translates $Q+k$ that intersect the boundary of A^jQ to the total number lying inside A^jQ converges to zero. To state this more precisely, let us define for each $j \geq 1$ the following finite subsets of Γ:

$$\begin{aligned} N_j &= \{k \in \Gamma : Q + k \subset A^jQ\}, \\ N_j^\circ &= \{k \in \Gamma : Q + k \subset (A^jQ)^\circ\}, \qquad (4.7) \\ N_j^\partial &= \{k \in \Gamma : Q + k \subset A^jQ \text{ and } (Q+k) \cap \partial(A^jQ) \neq \emptyset\}. \end{aligned}$$

By the remarks above, we have the following relationships:

$$\begin{aligned} A^jQ &= Q + N_j, \\ \#N_j &= m^j, \\ N_j^\circ \cup N_j^\partial &= N_j, \qquad (4.8) \\ N_j^\circ \cap N_j^\partial &= \emptyset. \end{aligned}$$

EXAMPLE 4.6. Consider the example of a uniform dilation of $\mathbf{R}^2$. That is, let $n = 2$, $A = 2I$, and $\Gamma = \mathbf{Z}^2$. Then $m = |\det(A)| = 4$. If we choose the digit set as $D = \{(0,0), (1,0), (0,1), (1,1)\}$, then the tile is the unit square $Q = [0,1]^2$. The dilated square $A^jQ = [0, 2^j]^2$ is tiled by 4^j translates of Q. It is easy to compute directly the number of translates of Q that touch the boundary of A^jQ. We find that

$$\begin{aligned} \#N_j^\circ &= (2^j - 2)^2 = 4^j - 2^{j+2} + 4, \\ \#N_j^\partial &= 4^j - (2^j - 2)^2 = 2^{j+2} - 4. \end{aligned}$$

Hence, the ratio $\#N_j^\circ/4^j$ approaches 1 as j increases, and the ratio $\#N_j^\partial/4^j$ approaches 0. ◊

The following result generalizes Example 4.6 to the case of an arbitrary dilation matrix, showing that $\#N_j^\circ$ is asymptotically on the order of m^j. This result can also be interpreted as an evaluation of the Beurling density of the lattice Γ.

LEMMA 4.7.

$$\lim_{j\to\infty} \frac{\#N_j^\circ}{m^j} = 1 \quad \textit{and} \quad \lim_{j\to\infty} \frac{\#N_j^\partial}{m^j} = 0. \qquad (4.9)$$

PROOF. For each $j \geq 1$, define

$$G_j = A^{-j}(Q + N_j^\circ) = \bigcup_{k \in N_j^\circ} A^{-j}(Q+k).$$

By definition, G_j is the union of all translates $A^{-j}(Q+k)$ that are contained within Q°. Each such "small tile" $A^{-j}(Q+k)$ is itself tiled by "smaller tiles" of the form $A^{-j-1}(Q+\ell)$. Those "smaller tiles" must be contained in Q° since they are contained in $A^{-j}(Q+k)$. Hence G_j is covered by translates $A^{-j-1}(Q+\ell)$ that are all contained in Q°, and therefore $G_j \subset G_{j+1}$ for all $j \geq 1$.

Since $G_j \subset Q^\circ$ by definition, we have $\cup G_j \subset Q^\circ$. We claim that, in fact, $\cup G_j = Q^\circ$. To see this, note that since A^{-1} is contractive and Q is bounded, the diameter of $A^{-j}Q$ converges to zero as j increases. Further, translates of $A^{-j}Q$ by elements of $A^{-j}\Gamma$ cover all of $\mathbf{R}^n$, i.e.,

$$A^{-j}Q + A^{-j}\Gamma = \bigcup_{k\in\Gamma} A^{-j}(Q+k) = \mathbf{R}^n.$$

Let $x \in Q^\circ$ be fixed. Then $\operatorname{dist}(x, \partial Q) > 0$. Hence, if j is large enough then there will exist some translate $A^{-j}(Q+k)$ that lies entirely within Q° and contains x. Hence $x \in G_j$ for that j. Thus $Q^\circ \subset \cup G_j$, as claimed.

Now, since the sets G_j are nested and their union is Q°, their measures must converge to the measure of Q°, i.e., $|G_j| \to |Q^\circ| = |Q|$. However, since $|\det(A^{-1})| = m^{-1}$ and since Γ-translates of Q have overlaps of measure zero, the Lebesgue measure of G_j is

$$|G_j| = |A^{-j}(Q+N_j^\circ)| = m^{-j}\,|Q+N_j^\circ| = m^{-j}\,|Q|\,\#N_j^\circ.$$

The first limit in (4.9) therefore follows. The second limit in (4.9) follows from the first limit and the relationships in (4.8). □

For later use, we now prove a technical lemma on the relationships among a set of tiles that cover an open ball B in $\mathbf{R}^n$. Let Ω be the minimal set of lattice points $k \in \Gamma$ such that $Q+k$ covers the ball B. The following lemma characterizes those translates $Q+\gamma$ of Q for which it is possible to translate $Q+\gamma$ by elements of Ω so that one translate $Q+\gamma+k$ with $k \in \Omega$ lies entirely within A^jQ and another translate $Q+\gamma+k'$ with $k' \in \Omega$ lies entirely outside of A^jQ (neglecting its boundary).

LEMMA 4.8. *Let B be an open ball in $\mathbf{R}^n$, and define*

$$\Omega = \{k \in \Gamma : (Q+k) \cap B \neq \emptyset\}. \tag{4.10}$$

Let $\gamma \in \Gamma$. If there exist k, $k' \in \Omega$ such that

$$Q+k+\gamma \subset A^jQ \quad \text{and} \quad Q+k'+\gamma \subset \mathbf{R}^n \setminus (A^jQ)^\circ, \tag{4.11}$$

then $\gamma \in N_j^\partial - \Omega = \{\ell - \omega : \ell \in N_j^\partial,\ \omega \in \Omega\}$.

PROOF. Note that Ω is finite and that $B \subset Q + \Omega$. Additionally, by definition of Ω,

$$(Q+k+\gamma)^\circ \cap (B+\gamma) \neq \emptyset \quad \text{and} \quad (Q+k'+\gamma)^\circ \cap (B+\gamma) \neq \emptyset.$$

Combined with (4.11), this implies that

$$(A^jQ)^\circ \cap (B+\gamma) \neq \emptyset \quad \text{and} \quad (\mathbf{R}^n \setminus A^jQ) \cap (B+\gamma) \neq \emptyset.$$

Since $B+\gamma$ is convex, there must therefore exist a line segment L entirely contained within $B+\gamma$ having one endpoint in $(A^jQ)^\circ$ and the other in $\mathbf{R}^n \setminus A^jQ$. Let $y \in L \cap \partial(A^jQ)$. Then there is some $\varepsilon > 0$ such that the open ball $B(y,\varepsilon)$ centered at y with radius ε lies entirely within $B+\gamma$.

Since $y \in \partial(A^jQ)$, there exists some $\ell \in N_j^\partial$ such that $y \in Q+\ell$. Then $B(y,\varepsilon) \cap (Q+\ell)^\circ \neq \emptyset$. Let $z \in B(y,\varepsilon) \cap (Q+\ell)^\circ$. Since z lies in the interior of

$Q + \ell$ and since translates of Q intersect only on their boundaries, we know that $Q + \ell$ is the *unique* lattice translate of Q that contains z. However,

$$z \in B(y, \varepsilon) \subset B + \gamma \subset Q + \Omega + \gamma,$$

so we must have $\ell = \omega + \gamma$ for some $\omega \in \Omega$. Consequently, $\gamma = \ell - \omega \in N_j^\partial - \Omega$, as desired. □

Our final lemma refines the estimates made in Lemma 4.7.

LEMMA 4.9. *If Ω is any finite subset of Γ, then*

$$\lim_{j\to\infty} \frac{\#\big(N_j^\circ \setminus ((N_j^\partial - \Omega) \cap N_j)\big)}{m^j} = 1.$$

PROOF. Note that

$$\#((N_j^\partial - \Omega) \cap N_j) \le \#(N_j^\partial - \Omega) \le \#N_j^\partial \cdot \#\Omega.$$

Hence,

$$\#N_j^\circ - \#N_j^\partial \cdot \#\Omega \le \#\big(N_j^\circ \setminus ((N_j^\partial - \Omega) \cap N_j)\big) \le \#N_j^\circ.$$

The result then follows from Lemma 4.7. □

Now we can give the proof of Theorem 4.4.

PROOF OF THEOREM 4.4. Suppose that the hypotheses of Theorem 4.4 are satisfied. Note that properties (P2) and (P5) are trivially satisfied by the definitions (4.2) and (4.3).

(b) Suppose that (P1) is satisfied. It then follows from (P2) and (P5) that

$$\{m^{1/2}\, \varphi_j(Ax - k)\}_{k\in\Gamma,\, j=1,\dots,r} \tag{4.12}$$

is an orthonormal basis for $\mathcal{V}_1$. By (P1) we have $\varphi_i \in \mathcal{V}_0 \subset \mathcal{V}_1$ for $i = 1, \dots, r$. The expansion of φ_i with respect to the orthonormal basis given in (4.12) is

$$\varphi_i(x) = m \sum_{j=1}^{r} \sum_{k\in\Gamma} \big\langle \varphi_i(x),\, \varphi_j(Ax - k)\big\rangle\, \varphi_j(Ax - k).$$

However, since φ_i has compact support, only finitely many terms in this series can be nonzero. Combining these equations for $i = 1, \dots, r$, we find that φ satisfies a refinement equation of the form (4.5).

Conversely, if φ satisfies a refinement equation of the form (4.5), then each translate $\varphi_i(x - k)$ is a finite linear combination of the functions $\varphi_j(Ax - \ell)$, each of which lies in $\mathcal{V}_1$. Since $\mathcal{V}_0$ is the closed linear span of the functions $\varphi_i(x - k)$, it follows that $\mathcal{V}_0 \subset \mathcal{V}_1$. Property (P1) then follows from this and the definition (4.3).

(a) As remarked above, the fact that (P2) and (P5) are satisfied is trivial. To show that (P3) holds, note first that $\{\varphi_i^{j,k}\}_{k\in\Gamma,\, i=1,\dots,r}$ is an orthonormal basis for the subspace $\mathcal{V}_j$. Therefore, if we let P_j denote the orthogonal projection of $L^2(\mathbf{R}^n)$ onto $\mathcal{V}_j$, then for each $g \in L^2(\mathbf{R}^n)$ we have

$$\|P_j g\|_{L^2}^2 = \sum_{i=1}^{r} \sum_{k\in\Gamma} |\langle g, \varphi_i^{j,k}\rangle|^2. \tag{4.13}$$

To prove (P3), it suffices to show that

$$\forall\, g \in L^2(\mathbf{R}^n), \qquad \lim_{j\to-\infty} \|P_j g\|_{L^2} \;=\; 0.$$

Moreover, it suffices to establish this limit for g contained in a complete subset of $L^2(\mathbf{R}^n)$, i.e., a subset whose finite linear span is dense in $L^2(\mathbf{R}^n)$. We will do this for the particular complete set given in Lemma 4.5, i.e., we will show that

$$\forall\, s \in \mathbf{Z}, \quad \forall\, \ell \in \Gamma, \qquad \lim_{j\to-\infty} \|P_j(\chi_Q^{s,\ell})\|_{L^2} \;=\; 0.$$

Fix any particular $s \in \mathbf{Z}$ and $\ell \in \Gamma$. Note that since $m = |\det(A)|$, we have for every $j \in \mathbf{Z}$ that

$$|A^{j-s}(Q+\ell)| \;=\; m^{j-s}\,|Q+\ell| \;=\; m^{j-s}\,|Q|.$$

Further, since A^{-1} is contractive, the sets $A^{j-s}(Q+\ell)$ for $j \le s$ are all contained inside a single compact set F. Also, the functions φ_i are compactly supported, so

$$K \;=\; \operatorname{supp}(\varphi) \;=\; \bigcup_{i=1}^{r} \operatorname{supp}(\varphi_i)$$

is compact. Therefore, there can be at most finitely many translates of K that intersect F, i.e., the set

$$J \;=\; \{k \in \Gamma : (K+k) \cap F \neq \emptyset\}$$

is finite. Applying (4.13) to $g = \chi_Q^{s,\ell}$ and using the facts above, we therefore compute that

$$\begin{aligned}
\|P_j(\chi_Q^{s,\ell})\|_{L^2}^2 &= \sum_{i=1}^{r}\sum_{k\in\Gamma} \left| \int m^{s/2}\,\chi_Q(A^s x - \ell)\, \overline{m^{j/2}\,\varphi_i(A^j x - k)}\, dx \right|^2 \\
&= \frac{1}{m^{j-s}} \sum_{i=1}^{r}\sum_{k\in\Gamma} \left| \int_{A^{j-s}(Q+\ell)} \varphi_i(x-k)\, dx \right|^2 \\
&\le \frac{|A^{j-s}(Q+\ell)|}{m^{j-s}} \sum_{i=1}^{r}\sum_{k\in J} \int_{A^{j-s}(Q+\ell)} |\varphi_i(x-k)|^2\, dx \\
&= |Q| \sum_{i=1}^{r}\sum_{k\in J} \int_{A^{j-s}(Q+\ell)} |\varphi_i(x-k)|^2\, dx, \qquad (4.14)
\end{aligned}$$

the inequality in this calculation following from Cauchy–Schwarz. Since each φ_i lies in $L^2(\mathbf{R}^n)$, since the sums in (4.14) are finite, and since the measure of $A^{j-s}(Q+\ell)$ converges to zero as $j \to -\infty$, it follows from (4.14) that $\|P_j(\chi_Q^{s,\ell})\|_{L^2}^2 \to 0$ as $j \to -\infty$.

(c) Note that if

$$\forall\, g \in L^2(\mathbf{R}^n), \qquad \lim_{j\to\infty} \|P_j g - g\|_{L^2} \;=\; 0, \tag{4.15}$$

then Property (P4) is satisfied. Further, if φ is refinable, then by part (a) we have $\mathcal{V}_j \subset \mathcal{V}_{j+1}$ for all $j \in \mathbf{Z}$, and therefore (4.15) is equivalent to Property (P4) when this additional assumption of refinability is satisfied. Therefore, to prove Theorem 4.4(c), it suffices to show that equations (4.15) and (4.6) are equivalent.

Let us first reformulate (4.15). By orthogonality we have $\|g - P_j g\|_{L^2}^2 = \|g\|_{L^2}^2 - \|P_j g\|_{L^2}^2$, so we can rewrite equation (4.15) as

$$\forall\, g \in L^2(\mathbf{R}^n), \quad \lim_{j\to\infty} \|P_j g\|_{L^2} \;=\; \|g\|_{L^2}. \tag{4.16}$$

As in the discussion for the proof of part (a), equation (4.16) is valid for all g if and only if it is valid for the specific functions $g = \chi_Q^{s,\ell}$ with $s \in \mathbf{Z}$ and $\ell \in \Gamma$. For the function χ_Q itself, we have from (4.13) that

$$\begin{aligned} \|P_j(\chi_Q)\|_{L^2}^2 &= \sum_{i=1}^{r} \sum_{k\in\Gamma} \left| \int \chi_Q(x)\, \overline{m^{j/2}\, \varphi_i(A^j x - k)}\, dx \right|^2 \\ &= \frac{1}{m^j} \sum_{i=1}^{r} \sum_{k\in\Gamma} \left| \int_{A^j Q} \varphi_i(x-k)\, dx \right|^2. \end{aligned} \tag{4.17}$$

For the function $\chi_Q^{s,\ell}(x) = m^{s/2}\, \chi_Q(A^s x - \ell)$, we have for $j \ge s$ that

$$\begin{aligned} \|P_j(\chi_Q^{s,\ell})\|_{L^2}^2 &= \sum_{i=1}^{r} \sum_{k\in\Gamma} \left| \int m^{s/2}\, \chi_Q(A^s x - \ell)\, \overline{m^{j/2}\, \varphi_i(A^j x - k)}\, dx \right|^2 \\ &= \frac{1}{m^{j-s}} \sum_{i=1}^{r} \sum_{k\in\Gamma} \left| \int_{A^{j-s} Q} \varphi_i(x - (k - A^{j-s}\ell))\, dx \right|^2 \\ &= \frac{1}{m^{j-s}} \sum_{i=1}^{r} \sum_{k\in\Gamma} \left| \int_{A^{j-s} Q} \varphi_i(x-k)\, dx \right|^2 \\ &= \|P_{j-s}\chi_Q\|_{L^2}^2. \end{aligned} \tag{4.18}$$

For the third equality in this calculation, we re-indexed the summation over k, using the fact that $A^{j-s}\ell \in \Gamma$ since $j - s \ge 0$. Comparing (4.17) and (4.18), we conclude that (4.16) is valid for all g if and only if it is valid for the single function $g = \chi_Q$. Further, since (4.15) and (4.16) are equivalent, we conclude that (4.15) is equivalent to the statement

$$\lim_{j\to\infty} \|P_j(\chi_Q)\|_{L^2}^2 \;=\; \|\chi_Q\|_{L^2}^2 \;=\; |Q|.$$

Hence, to prove that (4.15) is equivalent to (4.6), it suffices to show that

$$\lim_{j\to\infty} \|P_j(\chi_Q)\|_{L^2}^2 \;=\; \sum_{i=1}^{r} |\hat{\varphi}_i(0)|^2. \tag{4.19}$$

To estimate $\|P_j(\chi_Q)\|_{L^2}^2$, we will break the summation over Γ appearing in (4.17) into three regions related to the support of the functions φ_i, and then estimate the integrals corresponding to each of these regions in turn. The idea behind this division is that if $K = \operatorname{supp}(\varphi)$, then the first region should essentially contain only elements k of the lattice Γ such that $K + k$ is sure to lie in the interior of $A^j Q$, the second region should contain those k for which this translation will intersect the boundary of $A^j Q$, and the last region should be the complement of the first two.

More precisely, let B be any open ball in $\mathbf{R}^n$ which contains both Q and $K = \operatorname{supp}(\varphi)$, and define Ω by (4.10), i.e.,

$$\Omega \;=\; \{k \in \Gamma : (Q + k) \cap B \neq \emptyset\}.$$

Note that Ω is finite, that $B \subset Q + \Omega$, and that $0 \in \Omega$ since $Q \subset B$. Then for each $j \geq 1$, define:

$$\begin{aligned}
\Gamma_{1,j} &= N_j^\circ \setminus ((N_j^\partial - \Omega) \cap N_j), \\
\Gamma_{2,j} &= N_j^\partial - \Omega, \\
\Gamma_{3,j} &= \Gamma \setminus (\Gamma_{1,j} \cup \Gamma_{2,j}),
\end{aligned}$$

where the sets N_j, N_j° and N_j^∂ are as defined in (4.7). Note that for each j, the sets $\Gamma_{1,j}$, $\Gamma_{2,j}$, $\Gamma_{3,j}$ partition Γ. Further, by Lemmas 4.7 and 4.9 we have

$$\lim_{j\to\infty} \frac{\#\Gamma_{1,j}}{m^j} = 1 \qquad \text{and} \qquad \lim_{j\to\infty} \frac{\#\Gamma_{2,j}}{m^j} = 0. \tag{4.20}$$

Now define

$$R_{\nu,j} = \frac{1}{m^j} \sum_{i=1}^{r} \sum_{\gamma\in\Gamma_{\nu,j}} \left| \int_{A^jQ} \varphi_i(x-\gamma)\, dx \right|^2, \qquad \nu = 1,2,3.$$

Then, by (4.17),

$$\|P_j(\chi_Q)\|_{L^2}^2 = R_{1,j} + R_{2,j} + R_{3,j}.$$

Therefore, to prove (4.19), it suffices to prove the following three statements:

$$\lim_{j\to\infty} R_{1,j} = \sum_{i=1}^{r} |\hat{\varphi}_i(0)|^2, \qquad \lim_{j\to\infty} R_{2,j} = 0, \qquad \text{and} \qquad R_{3,j} = 0 \text{ for all } j.$$

($R_{3,j}$) Suppose that $R_{3,j} \neq 0$ for some j. Then $\int_{A^jQ} \varphi_i(x-\gamma)\,dx \neq 0$ for some $\gamma \in \Gamma_{3,j}$. This implies that $A^jQ \cap (K+\gamma)$ must have positive Lebesgue measure. Since $K \subset B \subset Q + \Omega$, and since the only translates of Q which intersect A^jQ in sets of positive measure are translates lying entirely within A^jQ, this implies that

$$Q + k + \gamma \subset A^jQ \qquad \text{for some } k \in \Omega. \tag{4.21}$$

Now, we have that $N_j^\partial \subset (N_j^\partial - \Omega) \cap N_j$ since $0 \in \Omega$ and $N_j^\partial \subset N_j$. Hence

$$N_j = N_j^\circ \cup N_j^\partial \subset \Gamma_{1,j} \cup \Gamma_{2,j}.$$

Since $\gamma \in \Gamma_{3,j}$, we must therefore have $\gamma \notin N_j$. By definition of $\mathbf{N}_j$, this implies that $Q+\gamma$ is not contained in A^jQ. Therefore $Q+\gamma \subset \mathbf{R}^n \setminus (A^jQ)^\circ$. Consequently,

$$Q + 0 + \gamma \subset \mathbf{R}^n \setminus (A^jQ)^\circ, \tag{4.22}$$

and since $0 \in \Omega$, it follows from Lemma 4.8 applied to (4.21) and (4.22) that $\gamma \in N_j^\partial - \Omega = \Gamma_{2,j}$. This is a contradiction, so we must have $R_{3,j} = 0$.

$(R_{2,j})$ Since φ_i is compactly supported and square-integrable, it is integrable. Therefore,

$$\begin{aligned} R_{2,j} &= \frac{1}{m^j} \sum_{i=1}^{r} \sum_{k \in \Gamma_{2,j}} \left| \int_{A^j Q} \varphi_i(x-k)\, dx \right|^2 \\ &\leq \frac{1}{m^j} \sum_{i=1}^{r} \sum_{k \in \Gamma_{2,j}} \left(\int_{\mathbf{R}^n} |\varphi_i(x)|\, dx \right)^2 \\ &= \frac{C \, \#\Gamma_{2,j}}{m^j}, \end{aligned}$$

so $R_{2,j} \to 0$ by (4.20).

$(R_{1,j})$ Suppose that $\gamma \in \Gamma_{1,j}$. Then $\gamma \in N_j^\circ$ and $\gamma \notin (N_j^\partial - \Omega)$. By definition of N_j°, we therefore have $Q + \gamma \subset (A^j Q)^\circ$. Since $0 \in \Omega$ and $Q + 0 + \gamma \subset (A^j Q)^\circ$, Lemma 4.8 implies that $Q + k + \gamma$ is not contained in $\mathbf{R}^n \setminus (A^j Q)^\circ$ for any $k \in \Omega$. Since $A^j Q$ is closed, this implies $Q + k + \gamma \subset A^j Q$ for all $k \in \Omega$. Hence

$$K + \gamma \;\subset\; B + \gamma \;\subset\; Q + \Omega + \gamma \;\subset\; A^j Q,$$

so

$$\int_{A^j Q} \varphi_i(x-\gamma)\, dx \;=\; \int_{\mathbf{R}^n} \varphi_i(x-\gamma)\, dx \;=\; \hat{\varphi}_i(0).$$

Therefore, by (4.20),

$$\begin{aligned} R_{1,j} &= \frac{1}{m^j} \sum_{i=1}^{r} \sum_{\gamma \in \Gamma_{1,j}} \left| \int_{A^j Q} \varphi_i(x-\gamma)\, dx \right|^2 \\ &= \frac{\#\Gamma_{1,j}}{m^j} \sum_{i=1}^{r} |\hat{\varphi}_i(0)|^2 \\ &\to \sum_{i=1}^{r} |\hat{\varphi}_i(0)|^2. \end{aligned}$$

□

4.2. Wavelets Associated with a Multiresolution Analysis

In this section we will assume that a multiresolution analysis of multiplicity r is given, and we will discuss the problem of the existence and construction of an orthonormal wavelet basis for $L^2(\mathbf{R}^n)$ associated to this MRA.

Assume that φ generates an MRA. Since $\mathcal{V}_0 \subset \mathcal{V}_1$, there exists a subspace $\mathcal{W}_0 \subset \mathcal{V}_1$ that is the orthogonal complement of $\mathcal{V}_0$ in $\mathcal{V}_1$. That is, all vectors in $\mathcal{V}_0$ are orthogonal to all vectors in $\mathcal{W}_0$, and $\mathcal{V}_1$ is the direct sum of $\mathcal{V}_0$ and $\mathcal{W}_0$, i.e., $\mathcal{V}_1 = \mathcal{W}_0 \oplus \mathcal{V}_0$. For each $j \in \mathbf{Z}$, let $\mathcal{W}_j$ be the subspace obtained from $\mathcal{W}_0$ analogously to how the subspace $\mathcal{V}_j$ is obtained from $\mathcal{V}_0$. That is, we let $\mathcal{W}_j$ consist of the dilation by A^j of all the functions in $\mathcal{W}_0$, i.e.,

$$\mathcal{W}_j \;=\; \{g(A^j x) : g \in \mathcal{W}_0\}.$$

Then we see immediately that $\mathcal{W}_j$ is the orthogonal complement of $\mathcal{V}_j$ in $\mathcal{V}_{j+1}$. In particular, $\mathcal{V}_{j+1} = \mathcal{W}_j \oplus \mathcal{V}_j$ for every $j \in \mathbf{Z}$. Iterating this fact, we have that if $j > 0$, then

$$\begin{aligned} \mathcal{V}_{j+1} &= \mathcal{W}_j \oplus \mathcal{V}_j \\ &= \mathcal{W}_j \oplus \mathcal{W}_{j-1} \oplus \mathcal{V}_{j-1} \\ &\vdots \\ &= \mathcal{W}_j \oplus \mathcal{W}_{j-1} \oplus \cdots \oplus \mathcal{W}_{-j} \oplus \mathcal{V}_{-j}. \end{aligned} \tag{4.23}$$

Since $\cup \mathcal{V}_j$ is dense in $L^2(\mathbf{R}^n)$ and $\cap \mathcal{V}_j = \{0\}$, if we let $j \to \infty$ in (4.23) we see that

$$L^2(\mathbf{R}^n) = \bigoplus_{j \in \mathbf{Z}} \mathcal{W}_j. \tag{4.24}$$

Furthermore, $\mathcal{W}_j \perp \mathcal{W}_k$ when $j \neq k$, so (4.24) is a decomposition of $L^2(\mathbf{R}^n)$ as a direct sum of orthogonal subspaces. Hence, if we can find an orthonormal basis $\mathcal{B}_j$ for each space $\mathcal{W}_j$, then $\cup \mathcal{B}_j$ will be an orthonormal basis for $L^2(\mathbf{R}^n)$. Moreover, since each space $\mathcal{W}_j$ is a dilation of $\mathcal{W}_0$, once we have an orthonormal basis $\mathcal{B}_0$ for $\mathcal{W}_0$, we can obtain an an orthonormal basis $\mathcal{B}_j$ for $\mathcal{W}_j$ simply by dilating all the elements of $\mathcal{B}_0$ by A^j and normalizing the results.

Hence our task reduces to finding an orthonormal basis for $\mathcal{W}_0$. We will seek a basis consisting of the lattice translates of a set of $m-1$ vector functions

$$\psi_\ell = (\psi_{\ell,1}, \dots, \psi_{\ell,r})^T \in L^2(\mathbf{R}^n, \mathbf{C}^r), \qquad \ell = 1, \dots, m-1.$$

That is, we seek an orthonormal basis $\mathcal{B}_0$ for $\mathcal{W}_0$ of the form

$$\mathcal{B}_0 = \{\psi_{\ell,i}(x-k)\}_{k\in\Gamma,\, i=1,\dots,r,\, \ell=1,\dots,m-1}.$$

This should be compared to the orthonormal basis for $\mathcal{V}_0$ given by (4.1). If such a basis can be found, then

$$\begin{aligned} \mathcal{B}_j &= \{m^{1/2}\psi_{\ell,i}(A^j x - k)\}_{k\in\Gamma,\, i=1,\dots,r,\, \ell=1,\dots,m-1} \\ &= \{\psi_{\ell,i}^{j,k}\}_{k\in\Gamma,\, i=1,\dots,r,\, \ell=1,\dots,m-1} \end{aligned}$$

will be an orthonormal basis for $\mathcal{W}_j$, and therefore

$$\bigcup_{j\in\mathbf{Z}} \mathcal{B}_j = \{\psi_{\ell,i}^{j,k}\}_{k\in\Gamma,\, i=1,\dots,r,\, \ell=1,\dots,m-1,\, j\in\mathbf{Z}} \tag{4.25}$$

will form the desired *orthonormal multiwavelet basis* for $L^2(\mathbf{R}^n)$. In this case, the $r(m-1)$ functions $\{\psi_{\ell,i} : i = 1,\dots,r,\ \ell = 1,\dots,m-1\}$ are the *multiwavelets* (or simply the *wavelets*) that generate this basis.

EXAMPLE 4.10. For motivation, let us review the one-dimensional, single-function case. Specifically, consider the case $n = 1$, $r = 1$, $A = 2$, $\Gamma = \mathbf{Z}$, and $D = \{0,1\}$. Assume that $\varphi \in L^2(\mathbf{R})$ generates an MRA for $L^2(\mathbf{R})$. Since $m = 2$, we seek a *single* wavelet $\psi \in L^2(\mathbf{R})$ such that $\{\psi(x-k)\}_{k\in\mathbf{Z}}$ forms an orthonormal basis for $\mathcal{W}_0$. Once this function is found, the orthonormal wavelet basis for $L^2(\mathbf{R})$ given by (4.25) will have the form $\{\psi^{j,k}\}_{j,k\in\mathbf{Z}}$.

The classical technique for finding this wavelet ψ is as follows. The vector scaling function φ satisfies a refinement equation of the form $\varphi(x) = \sum_{k\in\mathbf{Z}} c_k\, \varphi(2x-$

k). The symbol of this refinement equation is the 1-periodic function $m_0 \in L^2[0,1)$ defined by

$$m_0(\omega) \;=\; \frac{1}{2}\sum_{k\in\mathbf{Z}} c_k\, e^{-2\pi i k\omega}, \qquad \omega \in \mathbf{R}.$$

Note that if only finitely many coefficients c_k are nonzero, then m_0 is actually a trigonometric polynomial. This symbol m_0 is the unique function such that

$$\hat{\varphi}(2\omega) \;=\; m_0(\omega)\,\hat{\varphi}(\omega), \qquad \omega \in \mathbf{R}.$$

It can be shown that the function $\psi \in L^2(\mathbf{R})$ whose Fourier transform is defined by

$$\hat{\psi}(2\omega) \;=\; m_1(\omega)\,\hat{\varphi}(\omega), \qquad \omega \in \mathbf{R} \tag{4.26}$$

is a valid wavelet associated with this MRA if and only if $m_1(\omega) \in L^2[0,1)$ is a 1-periodic function such that the matrix

$$\mathcal{M}(\omega) \;=\; \big[m_i(\omega + \tfrac{j}{2})\big]_{i,j=0,1} \;=\; \begin{bmatrix} m_0(\omega) & m_0(\omega+\frac{1}{2}) \\ m_1(\omega) & m_1(\omega+\frac{1}{2}) \end{bmatrix} \tag{4.27}$$

is unitary for almost every ω. The success of one-dimensional wavelet theory is, in part, based on the fact that it is possible to *constructively* find such functions m_1. For example, we can take

$$m_1(\omega) \;=\; \frac{1}{2}\sum_{k\in\mathbf{Z}} (-1)^k\, \bar{c}_{1-k}\, e^{-2\pi i k\omega},$$

in which case (4.26) implies that

$$\psi(x) \;=\; \sum_{k\in\mathbf{Z}} (-1)^k\, \bar{c}_{1-k}\, \varphi(2x-k)$$

generates a wavelet basis for $L^2(\mathbf{R})$ [**Dau92**]. ◇

The results stated in Example 4.10 can be extended to the case of multivariate wavelets with arbitrary multiplicities and dilation matrices. We will state the relevant results here without proof, and for simplicity of notation we will restrict to the case where the lattice is $\Gamma = \mathbf{Z}^n$.

Let $\{\mathcal{V}_j\}_{j\in\mathbf{Z}}$ be an MRA of multiplicity r with associated vector scaling function $\varphi = (\varphi_1, \dots, \varphi_r)^T \in L^2(\mathbf{R}^n, \mathbf{C}^r)$. Then φ satisfies a refinement equation of the form

$$\varphi(x) \;=\; \sum_{k\in\Gamma} c_k\, \varphi(Ax-k)$$

for some matrices c_k in $\mathbf{C}^{r\times r}$. The symbol of this refinement equation is the 1-periodic matrix-valued function $M_0 \in L^2([0,1), \mathbf{C}^{r\times r})$ defined by

$$M_0(\omega) \;=\; \frac{1}{m}\sum_{k\in\Gamma} c_k\, e^{-2\pi i k\cdot\omega}, \qquad \omega \in \mathbf{R}^n.$$

This is the unique function satisfying

$$\hat{\varphi}(A^*\omega) \;=\; M_0(\omega)\,\hat{\varphi}(\omega), \qquad \omega \in \mathbf{R}^n.$$

Now suppose that $M_1, \dots, M_{m-1}$ are 1-periodic matrix-valued functions in $L^2([0,1), \mathbf{C}^{r\times r})$. Let us write these functions together with the function M_0 as

$$M_\ell(\omega) \;=\; \frac{1}{m}\sum_{k\in\Gamma} c_{\ell,k}\, e^{-2\pi i k\cdot\omega}, \qquad \ell = 0, \dots, m-1.$$

In particular, this means that $c_k = c_{0,k}$. Let $\psi_1, \dots, \psi_{m-1}$ be the vector functions in $L^2(\mathbf{R}^n, \mathbf{C}^r)$ whose Fourier transforms are defined by the formula

$$\hat{\psi}_\ell(A^*\omega) \;=\; M_\ell(\omega)\,\hat{\varphi}(\omega), \qquad \ell = 1, \dots, m-1.$$

We seek necessary and sufficient conditions on $M_1, \dots, M_{m-1}$ such that the lattice translates of $\{\psi_{\ell,i} : \ell = 1, \dots, m-1,\, i = 1, \dots, r\}$ will form an orthonormal basis for $\mathcal{W}_0$. These will be formulated in terms of a matrix $\mathcal{M}(\omega)$ analogous to the one defined in (4.27) for the one-dimensional case. Specifically, we choose a complete set of representatives $\gamma_0, \dots, \gamma_{m-1}$ of $\mathbf{Z}^n/A^*(\mathbf{Z}^n)$, and then define a matrix-valued function $\mathcal{M}(\omega)$ in block form by

$$\mathcal{M}(\omega) \;=\; [M_i(\omega + B^*\gamma_j)]_{i,j=0,\dots,m-1},$$

where $B = A^{-1}$. Note that $\mathcal{M}(\omega) \in (\mathbf{C}^{r\times r})^{m\times m}$ for each individual ω. We will say that $\mathcal{M}$ *is unitary a.e.* if for each i, $j = 0, \dots, m-1$ we have

$$\sum_{k=0}^{m-1} M_i(\omega + B^*\gamma_k)\, M_j^*(\omega + B^*\gamma_k) \;=\; \delta_{i,j} I_{r\times r} \quad \text{for a.e. } \omega \in \mathbf{R}^n.$$

Then we have the following theorem, whose proof is straightforward.

THEOREM 4.11. *Let $\{\mathcal{V}_j\}_{j\in\mathbf{Z}}$ be an MRA for $L^2(\mathbf{R}^n)$ of multiplicity r. Then, using the notation above, the following statements are equivalent.*

(a) $\{\psi_{\ell,i}(x-k)\}_{k\in\Gamma,\, i=1,\dots,r,\, \ell=1,\dots,m-1}$ *forms an orthonormal basis for* $\mathcal{W}_0$.

(b) $\mathcal{M}$ *is unitary a.e.*

(c) $\frac{1}{m}\sum_{k\in\Gamma} c_{i,k}\, c_{j,k-A\nu}^* \;=\; \delta_{0,\nu}\, \delta_{i,j}\, I_{r\times r}$ *for* $\nu \in \Gamma$ *and* i, $j = 0, \dots, m-1$.

Thus, once an MRA has been found, we can construct a wavelet basis for $L^2(\mathbf{R}^n)$ if we can construct a particular unitary matrix function $\mathcal{M}(\omega)$. For each ω, the matrix $\mathcal{M}(\omega)$ is of size $rm \times rm$, and the first r rows of this matrix are known. If the remaining rows can be completed so that $\mathcal{M}(\omega)$ is unitary a.e., then we can find the wavelets that generate the wavelet bases. Equivalently, we can try to solve the non-linear system of equations in (c).

The question of whether this completion can always be accomplished is a very difficult open question. It has been shown that if $(2m-2)r \geq n$ then $\mathcal{M}(\omega)$ can always be completed so as to be unitary a.e. However, even in this case it is usually difficult to complete the matrix in such a way that the associated wavelets have some specific properties. For example, it is not known whether, given a compactly supported vector scaling function, the matrix can be completed so that the wavelet is compactly supported. The existence of a wavelet set associated to an MRA for the case of a uniform dilation of $\mathbf{R}^n$ was proved by Gröchenig [**Grö87**], and is reproduced in [**Mey92**]. Results for a general dilation matrix A with multiplicity 1 are described in [**Woj97**]. The multivariable, multiwavelet case for a uniform dilation is studied in [**AK97**], cf. also [**Che97**].

CHAPTER 5

Examples

In this chapter we will show how the results of the previous chapters can be used to construct wavelet bases. We first apply them to a known example of a nonseparable orthonormal wavelet basis, and then use them to construct new examples of nonseparable orthonormal multiwavelet bases.

In Section 5.2 we will discuss the Kovačević–Vetterli scaling function. This is a known example of a nonseparable, continuous, compactly supported function that is refinable with respect to the quincunx dilation matrix

$$A = \begin{bmatrix} 1 & 1 \\ 1 & -1 \end{bmatrix}, \tag{5.1}$$

and which has orthonormal lattice translates. We use our techniques to obtain a numerical verification of the continuity of this scaling function.

In Section 5.3 we will construct new examples of nonseparable, continuous vector scaling functions with multiplicity $r = 2$ that are refinable with respect to the quincunx dilation A, have orthonormal lattice translates, and have accuracy equal or greater than the Kovačević–Vetterli scaling function. Additionally, we construct the multiwavelets corresponding to the MRA generated by these scaling functions, thus obtaining new multiwavelet bases for $L^2(\mathbf{R}^2)$.

Note that for the quincunx dilation A given in (5.1), we have $m = |\det(A)| = 2$. The corresponding lattice is $\Gamma = \mathbf{Z}^2$, and we fix the digit set as

$$D = \{(0,0),\, (1,0)\}.$$

With this choice, the tile Q is the parallelogram with vertices

$$\{(0,0),\, (1,0),\, (2,1),\, (1,1)\}.$$

This tile is pictured in Figure 2.1 in Chapter 2.

We will use the notation developed in previous chapters, applied now to the specific setting of the quincunx matrix. In particular, the techniques for characterizing the accuracy of a scaling function were presented in the general setting in Section 3.4. In the two-dimensional setting, the number of multi-indices of a given degree s is $d_s = s + 1$. We choose to order those multi-indices as $\{(s,0),\, (s-1,1),\, \dots,\, (0,s)\}$. With this ordering, the vector of all monomials of degree s is

$$X_{[s]}(x) = X_{[s]}(x_1, x_2) = \begin{bmatrix} x_1^s \\ x_1^{s-1}x_2 \\ \vdots \\ x_2^s \end{bmatrix}, \qquad x = (x_1, x_2) \in \mathbf{R}^2.$$

For $s = 0, 1, 2, 3$, the matrices $A_{[s]}$ introduced in Section 3.4 are given explicitly as

$$A_{[0]} = 1,$$

$$A_{[1]} = \begin{bmatrix} 1 & 1 \\ 1 & -1 \end{bmatrix},$$

$$A_{[2]} = \begin{bmatrix} 1 & 2 & 1 \\ 1 & 0 & -1 \\ 1 & -2 & 1 \end{bmatrix},$$

$$A_{[3]} = \begin{bmatrix} 1 & 3 & 3 & 1 \\ 1 & 1 & -1 & -1 \\ 1 & -1 & -1 & 1 \\ 1 & -3 & 3 & -1 \end{bmatrix}.$$

5.1. Numerical Estimates of the Joint Spectral Radius

The 2-JSR can often be computed exactly in terms of the spectral radius of a single matrix [**LM97**], [**Zho98**]. For other values of p, it can be difficult to compute the joint spectral radius exactly. In special cases, the uniform JSR can be computed easily from the eigenvalues of the matrices M_j. For example, if the M_j commute, or if they can be simultaneously triangularized or Hermitianized, then $\hat{\rho}_\infty(\mathcal{M})$ is the maximum of the absolute values of the eigenvalues of the M_j. However, this need not be true in general. It is true that if $\|\cdot\|$ is any matrix norm (i.e., a norm on $\mathbf{C}^{s\times s}$ which satisfies $\|AB\| \le \|A\|\,\|B\|$), and we define

$$\hat{\sigma}_{\infty,\ell} = \max_{\Pi\in\mathcal{P}_\ell} \rho(\Pi)^{1/\ell} \qquad \text{and} \qquad \hat{\rho}_{\infty,\ell} = \max_{\Pi\in\mathcal{P}_\ell} \|\Pi\|^{1/\ell},$$

then

$$\hat{\sigma}_{\infty,\ell} \le \hat{\rho}_\infty(\mathcal{M}) \le \hat{\rho}_{\infty,\ell} \qquad \text{for every } \ell. \tag{5.2}$$

This provides one means for numerically estimating a uniform JSR, although the number of matrix product computations involved grows exponentially with ℓ. However, the fact that the norm is submultiplicative implies that the following branch-and-bound algorithm, based on [**DL92**, Lemma 4.6], can be used for testing upper bound conjectures, cf. [**CH92**].

PROPOSITION 5.1. *Let $\mathcal{M} = \{M_1, \dots, M_m\}$ be a collection of $s \times s$ matrices, and let $\|\cdot\|$ be any matrix norm on $\mathbf{C}^{s\times s}$. Let $\delta > 0$ be given, and create a set $\mathcal{Q}$ of matrix products by implementing the following recursion m times, starting with $\Pi = M_i$ in turn for $i = 1, \dots, m$:*

- *If $\Pi = M_{\varepsilon_1} \cdots M_{\varepsilon_\ell}$ and $\|\Pi\|^{1/\ell} \le \delta$, then let $\Pi \in \mathcal{Q}$. Otherwise, repeat this step m times, replacing Π by each of ΠM_i in turn for $i = 1, \dots, m$.*

If this recursion terminates, then

$$\hat{\rho}(\mathcal{M}) \le \max_{\Pi\in\mathcal{Q}} \|\Pi\|^{1/\ell(\Pi)} \le \delta,$$

where $\ell(\Pi)$ is the length of the product Π. Moreover, this recursion must terminate if $\delta > \hat{\rho}(\mathcal{M})$, and cannot terminate if $\delta < \hat{\rho}(\mathcal{M})$.

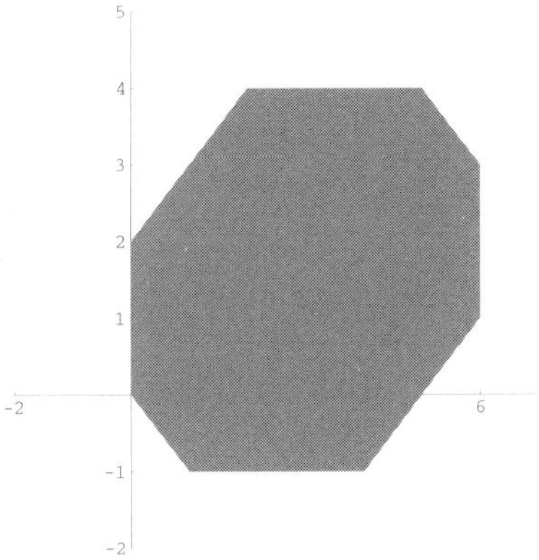

FIGURE 5.1. Attractor K_Λ.

This algorithm yields much better estimates with far less computation than the upper bound estimate given by (5.2), and often makes it possible to estimate the uniform JSR of quite large matrices with reasonable accuracy. Some analysis of the numerical accuracy of uniform JSR estimates is presented in [**Gri96**], and some methods for evaluating the exact uniform JSR of some types of collections $\mathcal{M}$ that cannot be simultaneously triangularized or Hermitianized can be found in [**Mae95**], [**BZ00**].

5.2. The Kovačević–Vetterli Scaling Function

The Kovačević–Vetterli (KV) scaling function was first constructed in [**KoV92**]. Until [**BW99**], it was the only known example of a nonseparable, continuous, compactly supported function $f\colon \mathbf{R}^2 \to \mathbf{C}$ that is refinable with respect to the quincunx matrix A and which has orthonormal lattice translates. The continuity of this function was conjectured in [**KoV92**] and was proved numerically in [**Vil94b**]. We will apply our techniques to obtain another numerical verification of the continuity of this scaling function.

The KV scaling function is the solution of the refinement equation

$$\varphi(x) = \sum_{k\in\Lambda} c_k\, \varphi(Ax-k), \qquad x \in \mathbf{R}^2, \tag{5.3}$$

for the following specific choices of Λ and c_k. The support of the coefficients is the following set of eight points in $\mathbf{Z}^2$:

$$\Lambda = \{(1,1),\ (2,1),\ (0,0),\ (1,0),\ (2,0),\ (3,0),\ (1,-1),\ (2,-1)\}.$$

The coefficients themselves are defined as follows. For $k \notin \Lambda$ let $c_k = 0$. Then define c_k for $k \in \Lambda$ by

$$[c_k]_{k\in\Lambda} = \nu \begin{bmatrix} & -a_1 & -a_0a_1 & \\ -a_2 & -a_0a_2 & -a_0 & 1 \\ & a_0a_1a_2 & -a_1a_2 & \end{bmatrix},$$

where the origin corresponds to the coefficient $-a_2$ and the scalar ν is chosen so that $\sum c_k = 2$. This gives a family of scaling functions, and the KV scaling function

$$c_{(1,1)} = \begin{bmatrix} -0.2626160679713805 & 0.4298190662052453 \\ 0.0005574439165755 & 0.2030577672486814 \end{bmatrix}$$

$$c_{(2,1)} = \begin{bmatrix} 0.0012426482475807 & -0.4949250389580165 \\ -0.0408719784870414 & -0.1920926795673339 \end{bmatrix}$$

$$c_{(0,0)} = \begin{bmatrix} 0.4558392979832848 & -0.1083434020875271 \\ 0.0706745015703368 & -0.0873302642653203 \end{bmatrix}$$

$$c_{(1,0)} = \begin{bmatrix} 1.0347430408665290 & -0.3333696001690321 \\ 0.0986292192873546 & -0.1130957347361869 \end{bmatrix}$$

$$c_{(2,0)} = \begin{bmatrix} 0.0217227622514353 & -0.1035804504304439 \\ -0.7252848187529292 & 0.3286159537916353 \end{bmatrix}$$

$$c_{(3,0)} = \begin{bmatrix} 0.0135277690777398 & 0.1053548733239501 \\ 0.1754378582933197 & -0.5539904957294699 \end{bmatrix}$$

$$c_{(1,-1)} = \begin{bmatrix} 0.0618708039296100 & -0.1876314721922069 \\ -0.2843099059597212 & 0.5949251108985801 \end{bmatrix}$$

$$c_{(2,-1)} = \begin{bmatrix} -0.0110715449521254 & -0.1958066410598297 \\ -0.1833149852352274 & 0.6670640666446144 \end{bmatrix}$$

$$v_{[0]} = \begin{bmatrix} 0.7920665605596084 & -0.6104347333198465 \end{bmatrix}$$

$$v_{[1]} = \begin{bmatrix} 1.3824676038808285 & -0.9905748274627678 \\ 0.7387595389423293 & -0.8523956367846645 \end{bmatrix}$$

TABLE 1. First set of scaling function coefficients.

corresponds to the specific choice

$$a_0 = a_1 = \sqrt{3}, \qquad a_2 = 2 - \sqrt{3}.$$

It follows from Proposition 3.3 that a compactly supported distributional solution φ to the refinement equation (5.3) exists. We will use the results of Chapter 3 to verify that this solution is in fact a continuous function, and to determine its accuracy. It is shown in [**Vil94b**] that lattice translates of φ are orthonormal.

First, we need to construct appropriate matrices T_d for $d \in D = \{(0,0), (1,0)\}$. With Λ as given above, the attractor K_Λ is the polygon with vertices

$$\{(0,0),\ (0,2),\ (2,4),\ (5,4),\ (6,3),\ (6,1),\ (4,-1),\ (1,-1)\}.$$

This polygon is pictured in Figure 5.1. By Proposition 2.2, the KV scaling function φ will be supported within K_Λ.

Let $\Omega \subset \mathbf{Z}^2$ be the set of 29 points with integer coordinates located within the polygon with vertices

$$\{(-1,-1),\ (-1,1),\ (1,3),\ (5,3),\ (5,1),\ (3,-1)\}.$$

This set Ω satisfies $K_\Lambda \subset Q + \Omega$, and, moreover, Ω is a minimal set with respect to this property. Then $T_{(0,0)}$ and $T_{(1,0)}$ are the two 29×29 matrices defined by (2.19),

$$c_{(1,1)} = \begin{bmatrix} -0.0591314043961276 & 0.4450003769119938 \\ 0.3321995579351313 & 0.0104446670717889 \end{bmatrix}$$

$$c_{(2,1)} = \begin{bmatrix} 0.1114102151429672 & 0.2254590843848077 \\ 0.1195224265431005 & -0.0593073985413613 \end{bmatrix}$$

$$c_{(0,0)} = \begin{bmatrix} -0.2940058981215972 & -0.4268371360660582 \\ -0.2507164723775025 & -0.0683481992678018 \end{bmatrix}$$

$$c_{(1,0)} = \begin{bmatrix} 0.6975221902082682 & 0.7758172235896774 \\ 0.1628550064232036 & 0.2857776144242880 \end{bmatrix}$$

$$c_{(2,0)} = \begin{bmatrix} -0.2453928326496505 & -0.0256314044863859 \\ 0.5286726350756744 & 0.8784799148003067 \end{bmatrix}$$

$$c_{(3,0)} = \begin{bmatrix} 0.3472507659119894 & -0.4193030011952396 \\ 0.0296262444198484 & 0.6570307353565332 \end{bmatrix}$$

$$c_{(1,-1)} = \begin{bmatrix} 0.0439292340612756 & -0.0386497896067917 \\ 0.0862583713567824 & 0.2000830628613316 \end{bmatrix}$$

$$c_{(2,-1)} = \begin{bmatrix} -0.0603512059818823 & 0.1176289221695265 \\ -0.3549334936719044 & -0.1969015485750640 \end{bmatrix}$$

$$v_{[0]} = \begin{bmatrix} -0.4088232319356361 & -0.9126135902060207 \end{bmatrix}$$

$$v_{[1]} = \begin{bmatrix} -1.6584856779704104 & -3.8822641730301039 \\ -0.5869518169744740 & 1.9243182275157703 \end{bmatrix}$$

$$v_{[2]} = \begin{bmatrix} -6.5634917083549151 & -16.5888850689201835 \\ -2.1313723063732684 & -8.2979351401696652 \\ -0.7721381710076713 & -4.0891841779249179 \end{bmatrix}$$

TABLE 2. Second set of scaling function coefficients.

i.e.,

$$T_{(0,0)} = [c_{Aj-k}]_{j,k\in\Omega} \qquad \text{and} \qquad T_{(1,0)} = [c_{Aj-k+(1,0)}]_{j,k\in\Omega}. \tag{5.4}$$

Now that the notation has been set, Theorem 3.12 and the remarks following imply that the accuracy of the KV scaling function is determined by the system of linear equations given in (3.18). All equations are given explicitly and exactly, and it is easy to check that the system can be solved when $\kappa = 2$, with solution

$$v_{[0]} = [v_{(0,0)}] = 1 \qquad \text{and} \qquad v_{[1]} = \begin{bmatrix} v_{(1,0)} \\ v_{(0,1)} \end{bmatrix} = \begin{bmatrix} (6+\sqrt{3})/2 \\ 3/2 \end{bmatrix}.$$

Furthermore, the system cannot be solved when $\kappa = 3$, so Theorem 3.12 implies that the KV scaling function has accuracy $\kappa = 2$, i.e., lattice translates of φ can reproduce exactly the constant and linear polynomials.

The vectors v_α given above directly determine the the polynomials y_α defined by (3.14), and these in turn determine directly the vectors e_α defined in (3.19).

$$d_{(1,1)} = \begin{bmatrix} -0.3241476668526600 & -0.6891609781760360 \\ 0.1112621003242060 & 0.3700696672377430 \end{bmatrix}$$

$$d_{(2,1)} = \begin{bmatrix} 0.2459217157892530 & 0.5440585582667560 \\ -0.1111644337484570 & -0.3131767059736590 \end{bmatrix}$$

$$d_{(0,0)} = \begin{bmatrix} 0.1342427066922970 & 0.3898586828404400 \\ 0.0008186524655252 & -0.1911338850170830 \end{bmatrix}$$

$$d_{(1,0)} = \begin{bmatrix} 0.6742751905644570 & 0.2690352672180090 \\ -0.1654413591601810 & -0.1668405479326910 \end{bmatrix}$$

$$d_{(2,0)} = \begin{bmatrix} 0.3023698075956100 & 0.1770706757135630 \\ 0.9358219858624430 & 0.2697411853489460 \end{bmatrix}$$

$$d_{(3,0)} = \begin{bmatrix} -0.1360696696971580 & -0.0664413982427058 \\ -0.3653701762481760 & 0.0578802392135144 \end{bmatrix}$$

$$d_{(1,-1)} = \begin{bmatrix} -0.0602396080362104 & 0.3209360059587710 \\ -0.0280398654007565 & 0.4891511174577440 \end{bmatrix}$$

$$d_{(2,-1)} = \begin{bmatrix} 0.1319145112771960 & 0.3110132133387610 \\ 0.3591016452044310 & 0.4405847665862590 \end{bmatrix}$$

TABLE 3. Wavelet coefficients corresponding to Table 2.

We now apply Theorems 3.4 and 3.17 to prove that φ is continuous. We apply the Gram–Schmidt procedure to $\{e_{(0,0)}, e_{(1,0)}, e_{(0,1)}\}$ to obtain an orthonormal basis $\{\tilde{e}_{(0,0)}, \tilde{e}_{(1,0)}, \tilde{e}_{(0,1)}\}$ for their span, which is the space called U_1 in the statement of Theorem 3.17. At the same time, the Gram–Schmidt procedure can be used to find an orthonormal basis $\mathcal{B}_E$ for the space $E_1 = \{e^*_{(0,0)}, e^*_{(1,0)}, e^*_{(0,1)}\}^\perp$. This yields an orthonormal basis for $\mathbf{C}^{29}$ of the form given by (3.21). In this basis, $T_{(0,0)}$ and $T_{(1,0)}$ have the form given in (3.22). Theorem 3.17 then implies that

$$\hat{\rho}_\infty(\{T_{(0,0)}|_{V_0}, T_{(1,0)}|_{V_0}\}) = \max\{\tfrac{1}{\sqrt{2}}, \hat{\rho}_\infty(C_0, C_1)\}, \tag{5.5}$$

where C_0 and C_1 are appropriate matrices of size 26×26. If this value is strictly less than 1, then Theorem 3.4 implies that φ is continuous.

To estimate the joint spectral radius in (5.5), we fix a norm and then implement the branch-and-bound algorithm of Proposition 5.1. We choose the norm to be the matrix norm induced by the ℓ^1 vector norm on $\mathbf{C}^{26}$. Then, following the recursion given in Proposition 5.1, a numerical computation of 1724 products of C_0 and C_1 yields the bound

$$\hat{\rho}_\infty(C_0, C_1) \leq 0.999713 < 1.$$

This therefore confirms the numerical proof of [**Vil94b**] that the KV scaling function exists and is continuous. A deeper computation of 42748 products of C_0, C_1, combined with the fact that $\rho(C_i) \leq \hat{\rho}_\infty(C_0, C_1)$, yields the numerical bounds $0.93407 \leq \hat{\rho}_\infty(C_0, C_1) \leq 0.94$.

5.3. Nonseparable Quincunx Multiwavelets

In this section we will present new examples of nonseparable, two-dimensional, compactly supported, continuous vector scaling functions of multiplicity 2 which are refinable with respect to the quincunx matrix A, have accuracy $\kappa = 2$ or 3, and have orthonormal lattice translates. The coefficients for these examples were provided to us by Anita Ruedin, see [**Rue02**] for related results. Ruedin used the characterization of higher-order accuracy developed in [**CHM98**], [**CHM00**] to construct candidate sets of coefficients. We will now give a numerical demonstration that these candidate vector scaling functions are in fact continuous, and we will construct the corresponding multiwavelets as well.

We use the same sets Λ, K_Λ, and Ω as were used in the definition and evaluation of the KV scaling function in Section 5.2. Let c_k for $k \in \Lambda$ be 2×2 matrices with unknown entries (a total of 32 unknowns). Suppose that there existed a solution φ to the refinement equation

$$\varphi(x) = \sum_{k\in\Lambda} c_k \, \varphi(Ax - k), \qquad x \in \mathbf{R}^2. \tag{5.6}$$

If this solution has orthonormal lattice translates, then necessarily

$$\forall j \in \mathbf{Z}^2, \quad \sum_{k\in\mathbf{Z}^2} c_k \, c_{k+Aj}^* = 2\delta_{j,0} I. \tag{5.7}$$

Taking into account the support of the coefficients, there are only 5 values of j for which (5.7) is nontrivial. This yields a set of 20 quadratic equations in the 32 unknown components of the c_k.

Now let

$$v_{[0]} = [v_{(0,0)}] \qquad \text{and} \qquad v_{[1]} = \begin{bmatrix} v_{(1,0)} \\ v_{(0,1)} \end{bmatrix},$$

where $v_{(0,0)}$, $v_{(1,0)}$, and $v_{(0,1)}$ are each unknown row vectors of length 2 (a total of 6 unknowns). If φ has accuracy $\kappa = 2$, then necessarily the equations in (3.18) must be satisfied, since φ has independent translates. This is a set of 8 linear equations in the variables that are the components of the c_k and the $v_{[s]}$.

Thus, if there exists a solution to the refinement equation (5.6) which has both orthonormal lattice translates and accuracy $\kappa = 2$, then a particular system of 28 linear and quadratics equations in 38 unknowns must be satisfied. Ruedin used a numerical optimization routine to produce sets of coefficients which satisfy each of these equations to within an accuracy of 3×10^{-13}. This set of coefficients is given in Table 1. A second set of coefficients, given in Table 2, satisfies to within an accuracy of 2×10^{-12} all of the equations specifying the necessary conditions for orthonormal lattice translates and accuracy $\kappa = 3$.

This information is not yet sufficient to imply that vector scaling functions with these properties do, in fact, exist. Proposition 3.3 does imply that compactly supported solutions to the refinement equations whose coefficients are given by Tables 1 and 2 do exist in at least the distributional sense. We will now demonstrate numerically that these solutions are continuous vector scaling functions. To do this, we apply Theorems 3.4 and 3.17, similarly to the verification that the KV scaling function is continuous.

Consider the values given in Table 1 first. The given vectors $v_{[0]}$ and $v_{[1]}$ directly determine the polynomials y_α defined in (3.14) and hence the vectors e_α defined in (3.19). The matrices $T_{(0,0)}$, $T_{(1,0)}$ are defined by the equations given in (5.4), except that the entries c_k are now 2×2 blocks. Hence each of these matrices has size 58×58. We make the change of basis to place $T_{(0,0)}$ and $T_{(1,0)}$ into the form given in (3.22). Theorem 3.17 then implies that

$$\hat{\rho}_\infty(\{T_{(0,0)}|_{V_0}, T_{(1,0)}|_{V_0}\}) \;=\; \max\{\tfrac{1}{\sqrt{2}},\, \hat{\rho}_\infty(C_0, C_1)\},$$

where C_0 and C_1 are appropriate matrices of size 55×55. We use the matrix norm induced by the ℓ^1 vector norm on $\mathbf{C}^{55}$. Then, following the recursion given in Proposition 5.1, a numerical computation of 1856 products of C_0 and C_1 yields the bound

$$\hat{\rho}_\infty(C_0, C_1) \;\le\; 0.999924 \;<\; 1.$$

Theorem 3.4 therefore implies that a continuous, compactly supported solution to this refinement equation does exist. A deeper computation of 226130 products of C_0, C_1, combined with the fact that $\rho(C_i) \le \hat{\rho}_\infty(C_0, C_1)$, yields the numerical bounds $0.714262 \le \hat{\rho}_\infty(C_0, C_1) \le 0.85$.

Theorem 3.4 also guarantees that the cascade algorithm converges. The vector scaling function $\varphi = (\varphi_1, \varphi_2)^{\mathrm{T}}$ is pictured in Figure 5.2 using a grid size of 1/16. The values at these grid points allow us to compute a Riemann sum approximation to the inner products $\langle \varphi_i(x-k), \varphi_j(x-\ell)\rangle$. These values equal $\delta_{ij}\delta_{k\ell}$ to within a precision of 8×10^{-3}, which we take as a numerical verification that lattice translates are orthonormal. Theorem 3.12 therefore implies that φ has accuracy $\kappa = 2$, i.e., translates of φ can reproduce constant and linear polynomials exactly. For example, we must have $\sum_k y_{(1,0)}(k)\, \varphi(x+k) = x_1$. In Figure 5.3 we show a partial sum of this series. Numerically, the full series $\sum_k y_{(1,0)}(k)\, f(x+k)$ equals x_1 to within an accuracy of 4×10^{-13}.

Since φ has orthonormal lattice translates and $\|\hat{\varphi}(0)\|^2 = 1$, it follows from Theorem 4.4 that φ generates a multiresolution analysis $\{\mathcal{V}_j\}$ for $L^2(\mathbf{R}^2)$. We can therefore use Theorem 4.11 to construct the corresponding multiwavelet basis for $L^2(\mathbf{R}^2)$. Specifically, since $m = 2$, we seek matrices $d_k = c_{1,k}$ for $k \in \Lambda$ so that the conditions in Theorem 4.11(c) are satisfied. The coefficients in Table 3 satisfy these conditions numerically to within an accuracy of 5×10^{-11}. The corresponding wavelets are shown in Figure 5.4.

Finally, consider the coefficients given in Table 2. These satisfy the necessary conditions for accuracy $\kappa = 3$. Because of the increased accuracy, we now have

$$\hat{\rho}_\infty(\{T_{(0,0)}|_{V_0}, T_{(1,0)}|_{V_0}\}) \;=\; \max\{\tfrac{1}{\sqrt{2}},\, \hat{\rho}_\infty(C_0, C_1)\}$$

with C_0 and C_1 being appropriate matrices of size 52×52. A numerical computation of 403850 products of C_0 and C_1, combined with the fact that $\rho(C_i) \le \hat{\rho}_\infty(C_0, C_1)$, yields the numerical bounds

$$0.91127 \;\le\; \hat{\rho}_\infty(C_0, C_1) \;\le\; 0.999999 \;<\; 1.$$

Hence a continuous vector scaling function exists, and is pictured in Figure 5.5. Translates of this vector scaling function can reproduce constant, linear, and quadratic polynomials exactly. In particular, we must have

$$\sum_k \bigl(y_{(2,0)}(k) + y_{(0,2)}(k)\bigr)\, f(x+k) \;=\; x_1^2 + x_2^2.$$

In Figure 5.6 we show a partial sum of this series. The corresponding wavelets can again be constructed by numerically solving the conditions presented in Theorem 4.11(c).

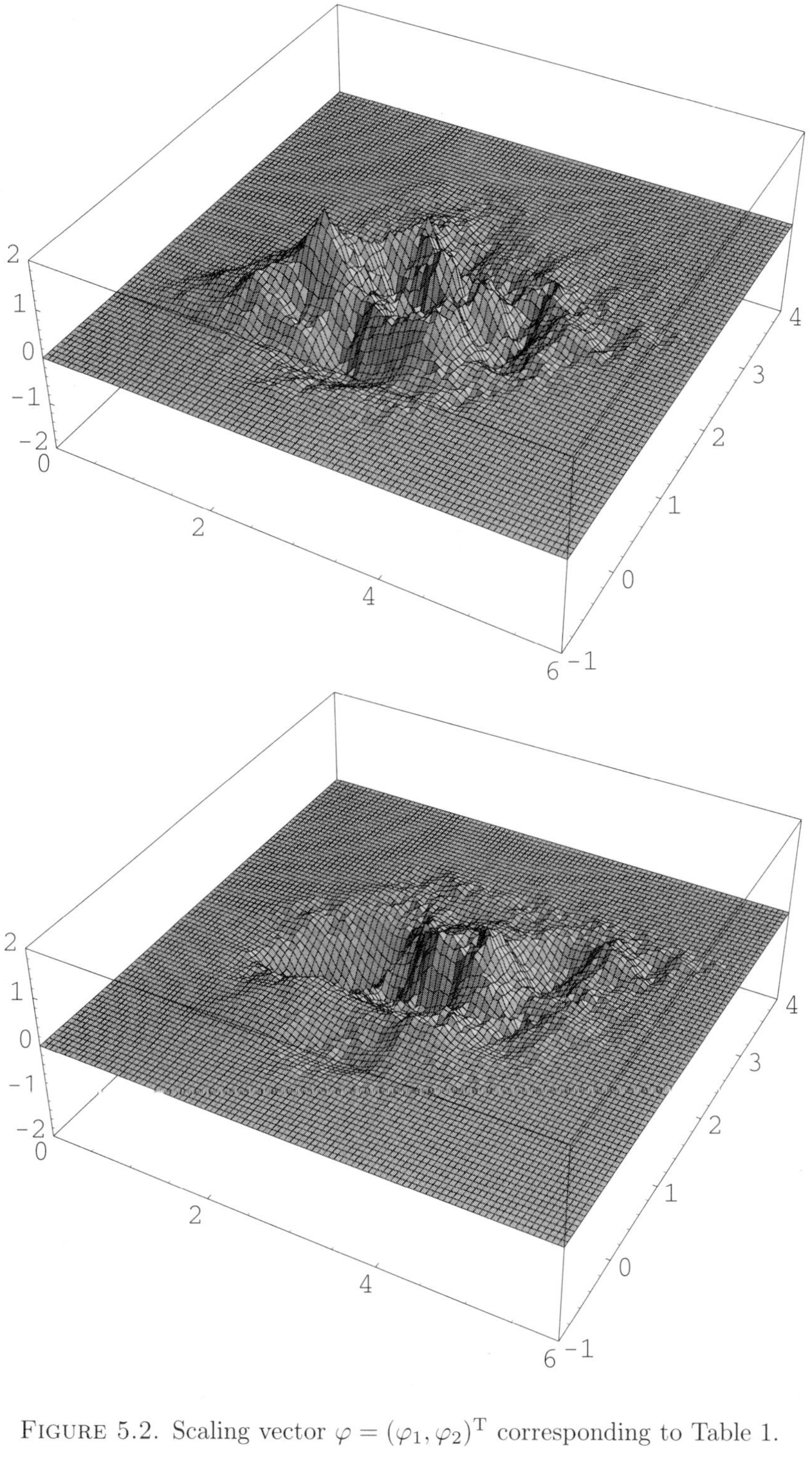

FIGURE 5.2. Scaling vector $\varphi = (\varphi_1, \varphi_2)^{\mathrm{T}}$ corresponding to Table 1.

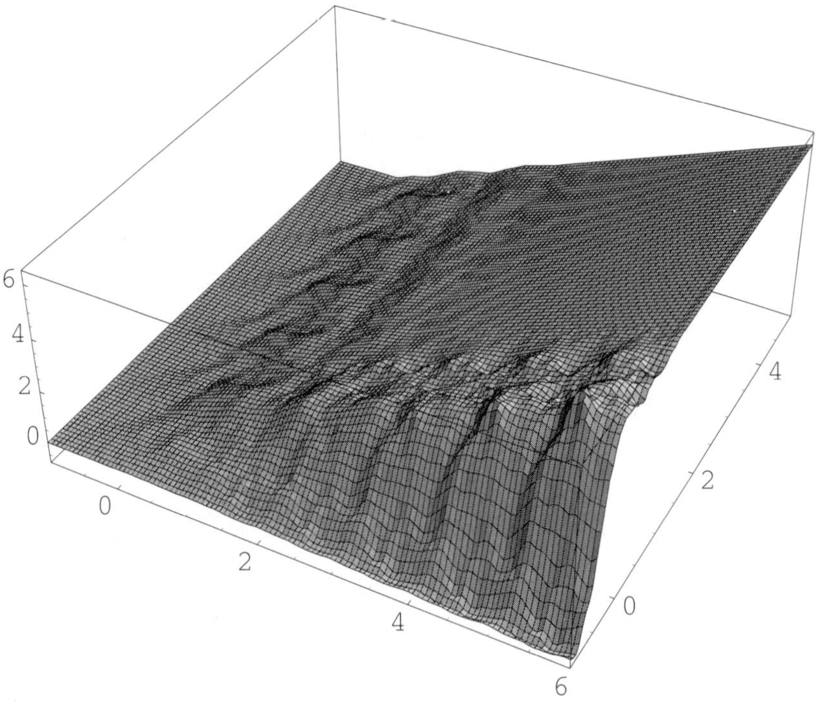

FIGURE 5.3. Partial sum of $\sum_k y_{(1,0)}(k)\,\varphi(x+k)$ for φ corresponding to Table 1.

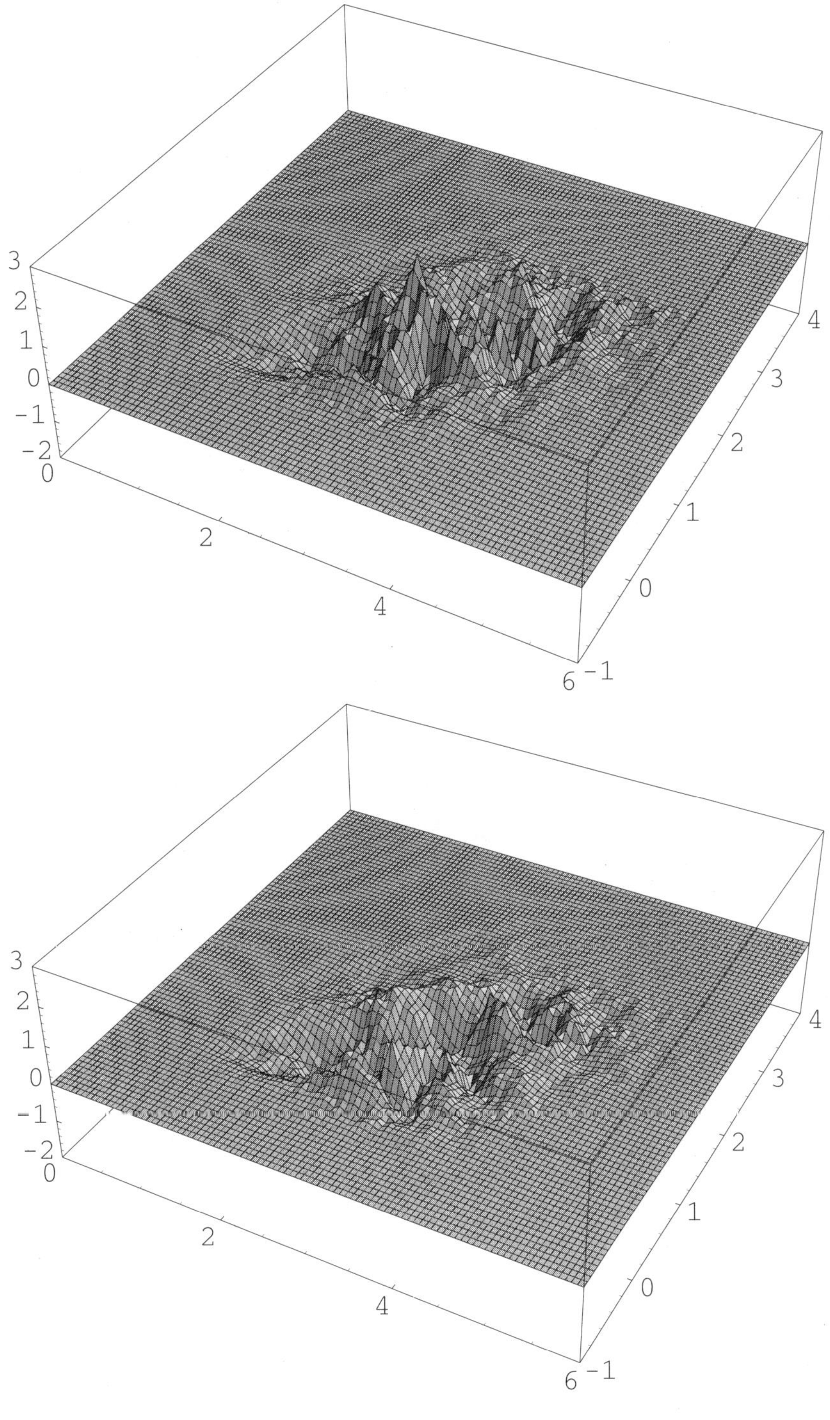

FIGURE 5.4. Wavelets $\psi = (\psi_1, \psi_2)^{\mathrm{T}}$ corresponding to Table 1.

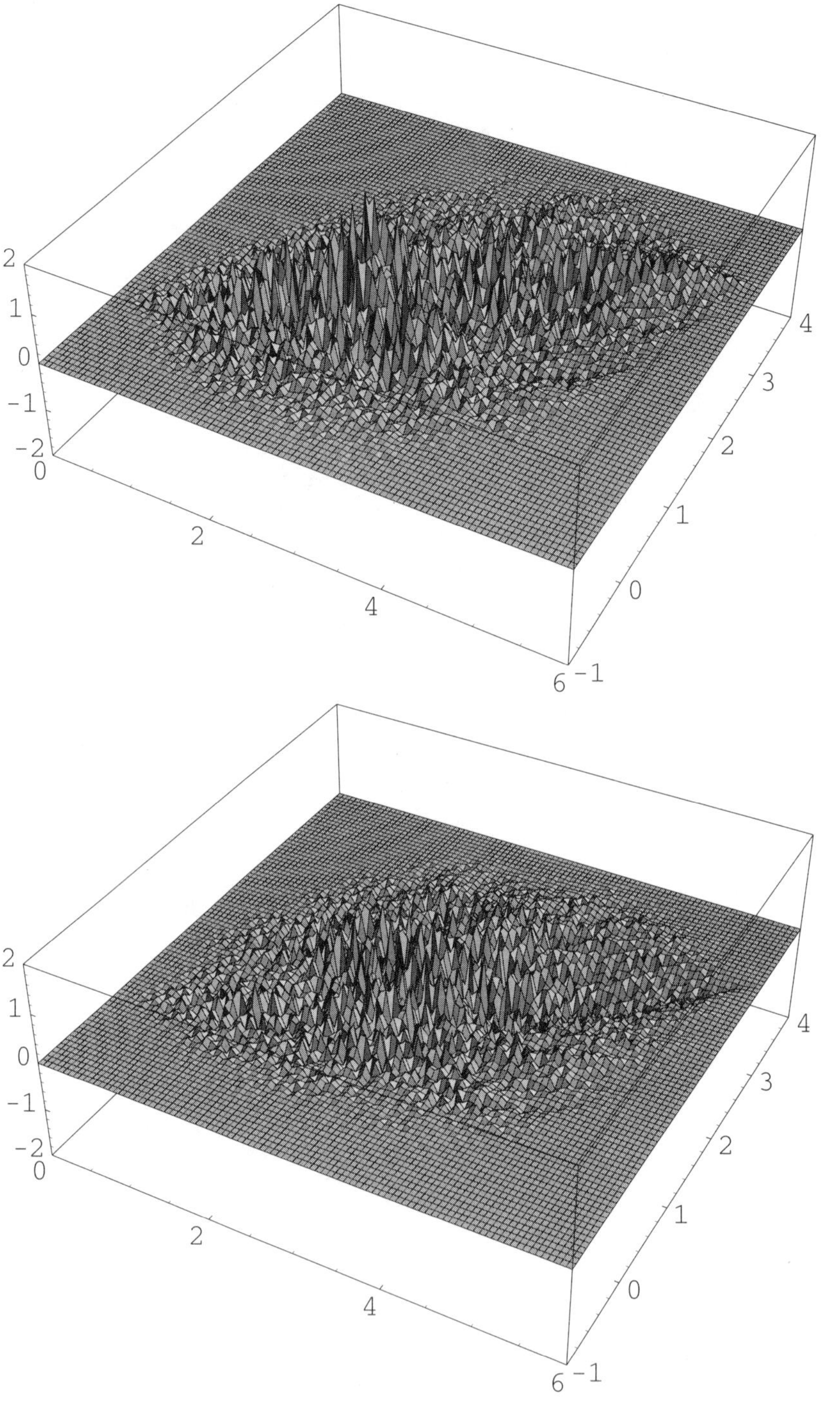

FIGURE 5.5. Scaling vector $\varphi = (\varphi_1, \varphi_2)^{\mathrm{T}}$ corresponding to Table 2.

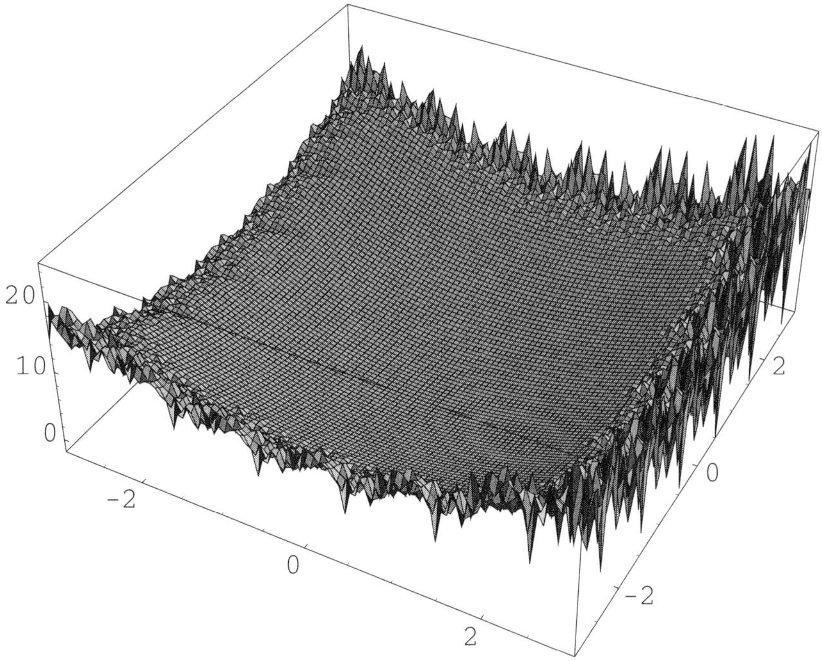

FIGURE 5.6. Partial sum of $\sum_k (y_{(2,0)}(k) + y_{(0,2)}(k))\, \varphi(x+k)$ for φ corresponding to Table 2.

Bibliography

[Alp93] B. K. Alpert, *A class of bases in L^2 for the sparse representation of integral operators*, SIAM J. Math. Anal. **24** (1993), 246–262.

[AK97] R. Ashino and M. Kametani, *A lemma on matrices and a construction of multiwavelets*, Math. Japan **45** (1997), 267–287.

[Aya99a] A. Ayache, *Construction of nonseparable dyadic compactly supported orthonormal wavelet bases for $L^2(\mathbf{R}^2)$ of arbitrarily high regularity*, Rev. Mat. Iberoamericana **15** (1999), 37–58.

[Aya99a] A. Ayache, *New filter banks that may generate nonseparable, compactly supported, orthonormal wavelet bases of arbitrarily high regularity*, preprint (1999).

[Ban91] C. Bandt, *Self-similar sets. V. Integer matrices and fractal tilings of $\mathbf{R}^n$*, Proc. Amer. Math. Soc. **112** (1991), 549–562.

[BW92] M. A. Berger and Y. Wang, *Bounded semi-groups of matrices*, Linear Algebra Appl. **166** (1992), 21–27.

[BW99] E. Belogay and Y. Wang, *Arbitrarily smooth orthogonal nonseparable wavelets in $\mathbf{R}^2$*, SIAM J. Math. Anal. **30** (1999), 678–697.

[BL01] J. J. Benedetto and M. Leon, *The construction of single wavelets in D-dimensions*, J. Geom. Anal. **11** (2001), 1–15.

[BDR94a] C. de Boor, R. De Vore, and A. Ron, *Approximation from shift-invariant subspaces of $L_2(\mathbf{R}^d)$*, Trans. Amer. Math. Soc. **341** (1994), 787–806.

[BDR94b] C. de Boor, R. De Vore, and A. Ron, *The structure of finitely generated shift-invariant subspaces of $L_2(\mathbf{R}^d)$*, J. Funct. Anal. **119** (1994), 37–78.

[BZ00] M. Bröker and X. Zhou, *Characterization of continuous, four-coefficient scaling functions via matrix spectral radius*, SIAM J. Matrix Anal. Appl. **22** (2000), 242–257.

[CHM98] C. Cabrelli, C. Heil, and U. Molter, *Accuracy of lattice translates of several multidimensional refinable functions*, J. Approx. Theory **95** (1998), 5–52.

[CHM99] C. A. Cabrelli, C. Heil, and U. M. Molter, *Polynomial reproduction by refinable functions*, Advances in Wavelets (Hong Kong, 1997), K.-S. Lau, ed., Springer–Verlag, Singapore, 1999, pp. 121–161.

[CHM00] C. Cabrelli, C. Heil, and U. Molter, *Accuracy of several multidimensional refinable distributions*, J. Fourier Anal. Appl. **6** (2000), 483–502.

[CM94] C. A. Cabrelli and U. M. Molter, *Generalization of a functional equation and its application to the construction of wavelets*, Preprint No. 78, Departamento de Matemática, Facultad de Cs. Exactas y Naturales, Universidad de Buenos Aires, 1994.

[CM99] C. A. Cabrelli and U. M. Molter, *Generalized self-similarity*, J. Math. Anal. Appl. **230** (1999), 251–260.

[Cal99] A. Calogero, *Wavelets on general lattices, associated with general expanding maps of $\mathbf{R}^n$*, Electron. Res. Announc. Amer. Math. Soc. **5** (1999), 1–10.

[Che97] D.-R. Chen, *On the existence and construction of orthonormal wavelets on $L^2(\mathbf{R}^s)$*, Proc. Amer. Math. Soc. **125** (1997), 2883–2889.

[Coh90] A. Cohen, *Ondelettes, analyses multirésolutions et filtres miroirs en quadrature*, Ann. Inst. H. Poincaré Anal. Non Linéaire **7** (1990), 439–459.

[CD93] A. Cohen and I. Daubechies, *Non-separable bidimensional wavelet bases*, Rev. Mat. Iberoamericana **9** (1993), 51–137.

[CDP97] A. Cohen, I. Daubechies, and G. Plonka, *Regularity of refinable function vectors*, J. Fourier Anal. Appl. **3** (1997), 295–324.

[CGV99] A. Cohen, K. Gröchenig, and L. F. Villemoes, *Regularity of multivariate refinable functions*, Constr. Approx. **15** (1999), 241–255.

[CH92] D. Colella and C. Heil, *The characterization of continuous, four-coefficient scaling functions and wavelets*, IEEE Trans. Inf. Th. **38** (1992), 876–881.

[CH94] D. Colella and C. Heil, *Characterizations of scaling functions: Continuous solutions*, SIAM J. Matrix Anal. Appl. **15** (1994), 496–518.

[DM97] S. Dahlke and P. Maass, *Interpolating refinable functions and wavelets for general scaling matrices*, Numer. Funct. Anal. Optim. **18** (1997), 521–539.

[DLS97] X. Dai, D. R. Larson, and D. M. Speegle, *Wavelet sets in* $\mathbf{R}^n$, J. Fourier Anal. Appl. **3** (1997), 451–456.

[Dau92] I. Daubechies, *Ten Lectures on Wavelets*, SIAM, Philadelphia, 1992.

[DL92] I. Daubechies and J. C. Lagarias, *Two-scale difference equations: II. Local regularity, infinite products of matrices and fractals*, SIAM J. Math. Anal. **23** (1992), 1031–1079.

[Der99] J. Derado, *Multivariate refinable interpolating functions*, Appl. Comput. Harmon. Anal. **7** (1999), 165–183.

[DGH95] G. C. Donovan, J. S. Geronimo, and D. P. Hardin, *A class of orthogonal multiresolution analyses in 2D*, Mathematical Methods for Curves and Surfaces, M. Dæhlen, T. Lyche, and L. L. Schumaker, eds., Vanderbilt University Press, Nashville, 1995, pp. 99–110.

[DGHM96] G. Donovan, J. S. Geronimo, D. P. Hardin, and P. R. Massopust, *Construction of orthogonal wavelets using fractal interpolation functions*, SIAM J. Math. Anal. **47** (1996), 1158–1192.

[Eir92] T. Eirola, *Sobolev characterization of solutions of dilation equations*, SIAM J. Math. Anal. **23** (1992), 1015–1030.

[GHM94] J. S. Geronimo, D. P. Hardin, and P. R. Massopust, *Fractal functions and wavelet expansions based on several scaling functions*, J. Approx. Th. **78** (1994), 373–401.

[GL94] T. N. T. Goodman, S. L. Lee, and W.-S. Tang, *Wavelets of multiplicity r*, Trans. Amer. Math. Soc. **342** (1994), 307–324.

[GLT93] T. N. T. Goodman, S. L. Lee, and W.-S. Tang, *Wavelets in wandering subspaces*, Trans. Amer. Math. Soc. **338** (1993), 639–654.

[Gri96] G. Gripenberg, *Computing the joint spectral radius*, Linear Algebra Appl. **234** (1996), 43–60.

[Grö87] K. Gröchenig, *Analyse multiéchelles et bases d'ondelettes*, C. R. Acad. Sci. Paris Sér. I Math. **305** (1987), 13–15.

[GM92] K. Gröchenig and W. R. Madych, *Multiresolution analysis, Haar bases, and self-similar tilings of* $\mathbf{R}^n$, IEEE Trans. Inform. Theory **38** (1992), 556–568.

[GR98] K. Gröchenig and A. Ron, *Tight compactly supported wavelet frames of arbitrarily high smoothness*, Proc. Amer. Math. Soc. **126** (1998), 1101–1107.

[Han97] B. Han, *On dual wavelet tight frames*, Appl. Comput. Harmon. Anal. **4** (1997), 380–413.

[HL97] W. He and M.-J. Lai, *Examples of bivariate nonseparable compactly supported orthonormal continuous wavelets*, Wavelet Applications in Signal and Image Processing IV, Proc. SPIE Vol. 2825, M. A. Unser et al., eds., SPIE, Bellingham, WA, 1997, pp. 303–314.

[HL98] W. He and M.-J. Lai, *Construction of bivariate compactly supported orthonormal multiwavelets with arbitrarily high regularity*, preprint, U. Georgia (1998).

[HL99] W. He and M.-J. Lai, *Construction of bivariate compactly supported biorthogonal box spline wavelets with arbitrarily high regularities*, Appl. Comput. Harmon. Anal. **6** (1999), 53–74.

[Hei94] C. Heil, *Some stability properties of wavelets and scaling functions*, Wavelets and Their Applications, J. S. Byrnes et al., eds., Kluwer Academic Publishers, Dordrecht, 1994, pp. 19–38.

[HC93] C. Heil and D. Colella, *Dilation equations and the smoothness of compactly supported wavelets*, Wavelets: Mathematics and Applications, J. J. Benedetto and M. W. Frazier, eds., CRC Press, Boca Raton, FL, 1993, pp. 161–200.

[HS95] C. Heil and G. Strang, *Continuity of the joint spectral radius*, Linear Algebra for Signal Processing, A. Bojanczyk and G. Cybenko, eds., IMA Vol. Math. Appl. **69**, Springer–Verlag, New York, 1995, pp. 51–61.

[HSS96] C. Heil, G. Strang, and V. Strela, *Approximation by translates of refinable functions*, Numerische Math. **73** (1996), 75–94.

[Hut81] J. Hutchinson, *Fractals and self-similarity*, Indiana Univ. Math. J. **30** (1981), 713–747.

[JRS99] H. Ji, S. D. Riemenschneider, and Z. Shen, *Multivariate compactly supported fundamental refinable functions, duals and biorthogonal wavelets*, Stud. Appl. Math. **102** (1999), 173–204.

[Jia95] R.-Q. Jia, *Subdivision schemes in L_p spaces*, Adv. Comput. Math. **3** (1995), 309–341.

[Jia99] R.-Q. Jia, *Characterization of smoothness of multivariate refinable functions in Sobolev spaces*, Trans. Amer. Math. Soc. **351** (1999), 4089–4112.

[Jng99] Q. Jiang, *Multivariate matrix refinable functions with arbitrary matrix dilation*, Trans. Amer. Math. Soc. **351** (1999), 2407–2438.

[KaV99] A. Karoui and R. Vaillancourt, *Nonseparable biorthogonal wavelet bases of $L^2(\mathbf{R}^n)$*, CRM Proc. Lecture Notes **18**, Amer. Math. Soc., Providence, RI, 1999, pp. 135–151.

[KoV92] J. Kovačević and M. Vetterli, *Nonseparable multidimensional perfect reconstruction filter banks and wavelet bases for $\mathbf{R}^n$*, IEEE Trans. Inform. Theory **38** (1992), 533–555.

[KoV95] J. Kovačević and M. Vetterli, *Nonseparable two- and three-dimensional wavelets*, IEEE Trans. Signal Proc. **43** (1995), 1269–1273.

[KS00] J. Kovačević and W. Sweldens, *Wavelet families of increasing order in arbitrary dimensions*, IEEE Trans. Image Proc. **9** (2000¡ 480–396.

[LagW95a] J. C. Lagarias and Y. Wang, *Haar type orthonormal wavelet bases in $\mathbf{R}^2$*, J. Fourier Anal. Appl. **2** (1995), 1–14.

[LagW95b] J. C. Lagarias and Y. Wang, *The finiteness conjecture for the generalized spectral radius of a set of matrices*, Linear Algebra Appl. **214** (1995), 17–42.

[LagW96] J. C. Lagarias and Y. Wang, *Haar bases for $L^2(\mathbf{R}^n)$ and algebraic number theory*, J. Number Theory **57** (1996), 181–197.

[LagW97] J. C. Lagarias and Y. Wang, *Integral self-affine tiles in $\mathbf{R}^n$. II. Lattice tilings*, J. Fourier Anal. Appl. **3** (1997), 83–102.

[LagW99] *Corrigendum and addendum to: Haar bases for $L^2(\mathbf{R}^n)$ and algebraic number theory*, J. Number Theory **76** (1999), 330–336.

[LM97] K.-S. Lau and M.-F. Ma, *The regularity of L^2-scaling functions*, Asian J. Math. **1** (1997), 272–292.

[LauW95] K.-S. Lau and J. Wang, *Characterization of L^p-solutions for the two-scale dilation equation*, SIAM J. Math. Anal. **26** (1995), 1018–1046.

[Mae95] M. Maesumi, *Optimum ball for joint spectral radius: an example from four-coefficient MRA*, Approximation Theory VIII, Vol. 2, C. K. Chui and L. L. Schumaker, eds., World Scientific, Singapore, 1995, pp. 267–274.

[Mey92] Y. Meyer, *Wavelets and Operators*, Cambridge University Press, Cambridge, 1992.

[MP89] C. A. Micchelli and H. Prautzsch, *Uniform refinement of curves*, Linear Algebra Appl. **114/115** (1989), 841–870.

[MS97] C. A. Micchelli and T. Sauer, *Regularity of multiwavelets*, Adv. Comput. Math. **7** (1997), 455–545.

[Plo97] G. Plonka, *Approximation order of shift-invariant subspaces of $L^2(\mathbf{R})$ generated by refinable function vectors*, Constr. Approx. **13** (1997), 221–244.

[Pot97] A. Potiopa, *A problem of Lagarias and Wang*, Master's Thesis, Siedlce University, Siedlce, Poland, 1997 (Polish).

[RS60] G. C. Rota and G. Strang, *A note on the joint spectral radius*, Kon. Nederl. Akad. Wet. Proc. A **63** (1960), 379–381.

[Rue02] A. M. C. Ruedin, *Construction of nonseparable multiwavelets for nonlinear image compression*, EURASIP J. Appl. Signal Process. **2002** (2002), 73–79.

[Sun91] Q. Sun, *Two-scale difference equations: Local and global linear independence*, unpublished manuscript (1991).

[Vil92] L. F. Villemoes, *Energy moments in time and frequency for two-scale difference equation solutions and wavelets*, SIAM J. Math. Anal. **23** (1992), 1519–1543.

[Vil94a] L. F. Villemoes, *Wavelet analysis of refinement equations*, SIAM J. Math. Anal. **25** (1994), 1433–1460.

[Vil94b] L. F. Villemoes, *Continuity of nonseparable quincunx wavelets*, Appl. Comput. Harmon. Anal. **1** (1994), 180–187.

[Wan95] Y. Wang, *Two-scale dilation equations and the cascade algorithm*, Random Comput. Dynam. **3** (1995), 289–307.

[Wan96] Y. Wang, *Two-scale dilation equations and the mean spectral radius*, Random Comput. Dynam. **4** (1996), 49–72.

[Woj97] P. Wojtaszczyk, *A Mathematical Introduction to Wavelets*, Cambridge University Press, Cambridge, 1997.

[Zho98] D.-X. Zhou, *The p-norm joint spectral radius for even integers*, Methods Appl. Anal. **5** (1998), 39–54.

APPENDIX A

Index of Symbols

Symbol	Meaning	Discussion
B^{T}	transpose of a matrix B	2.1
B^*	Hermitian of a matrix B	2.1
B_ε	$= \{x \in \mathbf{R}^n : \mathrm{dist}(B, x) < \varepsilon\}$	4.2.1
$\#F$	cardinality of a set F	2.1
E°	interior of $E \subset \mathbf{R}^n$	2.1
∂E	boundary of $E \subset \mathbf{R}^n$	2.1
$\overline{E}$	closure of $E \subset \mathbf{R}^n$	2.1
$\|E\|$	Lebesgue measure of $E \subset \mathbf{R}^n$	2.1
$f^{(i)}$	iterate in the Cascade algorithm	2.1
$\hat{f}$	Fourier transform of f	2.1
$g^{j,k}$	$= m^{j/2} g(A^j x - k)$	4.1
$.\varepsilon_1 \varepsilon_2 \dots$	A-nary expansion	2.2.1
A	dilation matrix	1.1, 2.1
$A_{[s]}$	matrix related to dilation of $X_{[s]}$ by A	3.4
$B(x, \varepsilon)$	open ball centered at x with radius ε	2.1
$\mathbf{C}^{J \times K}$	space of $J \times K$ complex matrices	2.1
c_k	coefficient in the refinement equation	1.1, 2.1
D	$= \{d_1, \dots, d_m\}$, digit set associated with A	2.2
d_s	number of multi-indices of degree s	3.4
e_α	$= (y_\alpha(k))_{k \in \Omega}$	3.5
E_s	$= \mathrm{span}\{e_\alpha^* : 0 \le \|\alpha\| \le s\}^\perp$	3.5
$\mathcal{H}(\mathbf{R}^n)$	Hausdorff space	2.2.1
$h(\cdot, \cdot)$	Hausdorff metric	2.2.1
K_H	attractor of IFS $\{w_k\}_{k \in H}$	2.2.1
L	$= [c_{Ai-j}]_{i,j \in \Gamma}$	3.4
$L^p(X)$	space of p-integrable functions $g \colon X \to \mathbf{C}$	2.1
$L^p(X, Y)$	space of p-integrable functions $g \colon X \to Y$	2.1
m	$= \|\det(A)\|$	2.2
$m_0(\omega)$	symbol of the refinement equation	4.2
n	dimension of domain of scaling function	1.1, 2.1
P	fundamental domain for $\Gamma / A(\Gamma)$	2.2
P_j	orthogonal projection of $L^2(\mathbf{R}^n)$ onto $\mathcal{V}_j$	4.1
Q	$= K_D$, tile associated with A and D	2.2.2
$\tilde{Q}$	subset of Q, tiles without overlaps	2.2.3
Q_i	"disjointization" of $w_{d_i}(Q)$	2.3
$Q_{[s,t]}$	a matrix of polynomials related to accuracy	3.4

r	multiplicity of scaling function	1.1, 2.1
S	refinement operator	2.1
$\mathrm{supp}(f)$	support of f	2.1
T	matrix version of refinement operator S	2.3
T_d	$= [c_{Aj-k+d}]_{j,k\in\Omega}$	2.3
U_s	$= \mathrm{span}\{e_\alpha : 0 \leq \lvert\alpha\rvert \leq s\}$	3.5
$\mathcal{V}_j$	subspaces in a multiresolution analysis	4.1
v_α	row vectors related to accuracy	3.4
$v_{[s]}$	$= [v_\alpha]_{\lvert\alpha\rvert=s}$	3.4
$\mathcal{W}_j$	orthogonal complement of $\mathcal{V}_j$ in $\mathcal{V}_{j+1}$	4.1
w_k	affine map, $w_k(x) = A^{-1}(x+k)$	2.2.1
w_H	$w_H(B) = \cup_{k\in H} w_k(B)$	2.2.1
$X_{[s]}$	vector of all monomials of degree s	3.4
y_α	a vector of polynomials related to accuracy	3.4
$y_{[s]}$	$= [y_\alpha]_{\lvert\alpha\rvert=s}$	3.4
$Y_{[s]}$	$= \big(y_{[()}x + k)\big)_{k\in\Gamma}$	3.4
Δ	$= \frac{1}{m}\sum c_k$	3.2
$\delta_{i,j}$	Kronecker delta	2.1
Γ	lattice in $\mathbf{R}^n$ invariant under A	1.1, 2.1
Γ_d	cosets of $A(\Gamma)$	2.2
γ_0	unique element of $\tilde{Q}\cap\Gamma$	3.6
γ_i	generators of the lattice Γ	2.2
κ	accuracy	1.1, 3.2
Λ	support of coefficients in refinement equation	1.1, 2.1
Λ'	$= \Lambda - D$	2.2.3, 3.5
Φg	folding of g	2.3
$\rho(M)$	spectral radius of matrix M	2.4
$\hat{\rho}_p(\mathcal{M})$	p-norm joint spectral radius of set of matrices $\mathcal{M}$	2.4
τ	analogue of $2x$ mod 1 map	2.3
Ω	subset of Γ such that $K_\Lambda \subset Q + \Omega$	2.2.3
χ_E	characteristic function of a set E	2.1

Editorial Information

To be published in the *Memoirs*, a paper must be correct, new, nontrivial, and significant. Further, it must be well written and of interest to a substantial number of mathematicians. Piecemeal results, such as an inconclusive step toward an unproved major theorem or a minor variation on a known result, are in general not acceptable for publication. Papers appearing in *Memoirs* are generally longer than those appearing in *Transactions*, which shares the same editorial committee.

As of March 1, 2004, the backlog for this journal was approximately 4 volumes. This estimate is the result of dividing the number of manuscripts for this journal in the Providence office that have not yet gone to the printer on the above date by the average number of monographs per volume over the previous twelve months, reduced by the number of volumes published in four months (the time necessary for preparing a volume for the printer). (There are 6 volumes per year, each containing at least 4 numbers.)

A Consent to Publish and Copyright Agreement is required before a paper will be published in the *Memoirs*. After a paper is accepted for publication, the Providence office will send a Consent to Publish and Copyright Agreement to all authors of the paper. By submitting a paper to the *Memoirs*, authors certify that the results have not been submitted to nor are they under consideration for publication by another journal, conference proceedings, or similar publication.

Information for Authors

Memoirs are printed from camera copy fully prepared by the author. This means that the finished book will look exactly like the copy submitted.

The paper must contain a *descriptive title* and an *abstract* that summarizes the article in language suitable for workers in the general field (algebra, analysis, etc.). The *descriptive title* should be short, but informative; useless or vague phrases such as "some remarks about" or "concerning" should be avoided. The *abstract* should be at least one complete sentence, and at most 300 words. Included with the footnotes to the paper should be the 2000 *Mathematics Subject Classification* representing the primary and secondary subjects of the article. The classifications are accessible from `www.ams.org/msc/`. The list of classifications is also available in print starting with the 1999 annual index of *Mathematical Reviews*. The Mathematics Subject Classification footnote may be followed by a list of *key words and phrases* describing the subject matter of the article and taken from it. Journal abbreviations used in bibliographies are listed in the latest *Mathematical Reviews* annual index. The series abbreviations are also accessible from `www.ams.org/publications/`. To help in preparing and verifying references, the AMS offers MR Lookup, a Reference Tool for Linking, at `www.ams.org/mrlookup/`. When the manuscript is submitted, authors should supply the editor with electronic addresses if available. These will be printed after the postal address at the end of the article.

Electronically prepared manuscripts. The AMS encourages electronically prepared manuscripts, with a strong preference for $\mathcal{A}\mathcal{M}\mathcal{S}$-LaTeX. To this end, the Society has prepared $\mathcal{A}\mathcal{M}\mathcal{S}$-LaTeX author packages for each AMS publication. Author packages include instructions for preparing electronic manuscripts, the *AMS Author Handbook*, samples, and a style file that generates the particular design specifications of that publication series. Though $\mathcal{A}\mathcal{M}\mathcal{S}$-LaTeX is the highly preferred format of TeX, author packages are also available in $\mathcal{A}\mathcal{M}\mathcal{S}$-TeX.

Authors may retrieve an author package from e-MATH starting from `www.ams.org/tex/` or via FTP to `ftp.ams.org` (login as `anonymous`, enter username as password, and type `cd pub/author-info`). The *AMS Author Handbook* and the *Instruction Manual* are available in PDF format following the author packages link from `www.ams.org/tex/`. The author package can be obtained free of charge by sending email to `pub@ams.org` (Internet) or from the Publication Division, American Mathematical Society, 201 Charles St., Providence, RI 02904, USA. When requesting an author package, please specify AMS-LaTeX or AMS-TeX, Macintosh or IBM (3.5) format, and the publication in which your paper will appear. Please be sure to include your complete mailing address.

Sending electronic files. After acceptance, the source file(s) should be sent to the Providence office (this includes any TeX source file, any graphics files, and the DVI or PostScript file).

Before sending the source file, be sure you have proofread your paper carefully. The files you send must be the EXACT files used to generate the proof copy that was accepted for publication. For all publications, authors are required to send a printed copy of their paper, which exactly matches the copy approved for publication, along with any graphics that will appear in the paper.

TeX files may be submitted by email, FTP, or on diskette. The DVI file(s) and PostScript files should be submitted only by FTP or on diskette unless they are encoded properly to submit through email. (DVI files are binary and PostScript files tend to be very large.)

Electronically prepared manuscripts can be sent via email to `pub-submit@ams.org` (Internet). The subject line of the message should include the publication code to identify it as a Memoir. TeX source files, DVI files, and PostScript files can be transferred over the Internet by FTP to the Internet node `e-math.ams.org` (130.44.1.100).

Electronic graphics. Comprehensive instructions on preparing graphics are available at `www.ams.org/jourhtml/graphics.html`. A few of the major requirements are given here.

Submit files for graphics as EPS (Encapsulated PostScript) files. This includes graphics originated via a graphics application as well as scanned photographs or other computer-generated images. If this is not possible, TIFF files are acceptable as long as they can be opened in Adobe Photoshop or Illustrator. No matter what method was used to produce the graphic, it is necessary to provide a paper copy to the AMS.

Authors using graphics packages for the creation of electronic art should also avoid the use of any lines thinner than 0.5 points in width. Many graphics packages allow the user to specify a "hairline" for a very thin line. Hairlines often look acceptable when proofed on a typical laser printer. However, when produced on a high-resolution laser imagesetter, hairlines become nearly invisible and will be lost entirely in the final printing process.

Screens should be set to values between 15% and 85%. Screens which fall outside of this range are too light or too dark to print correctly. Variations of screens within a graphic should be no less than 10%.

Inquiries. Any inquiries concerning a paper that has been accepted for publication should be sent directly to the Electronic Prepress Department, American Mathematical Society, 201 Charles St., Providence, RI 02904, USA.

Editors

This journal is designed particularly for long research papers, normally at least 80 pages in length, and groups of cognate papers in pure and applied mathematics. Papers intended for publication in the *Memoirs* should be addressed to one of the following editors. In principle the Memoirs welcomes electronic submissions, and some of the editors, those whose names appear below with an asterisk (*), have indicated that they prefer them. However, editors reserve the right to request hard copies after papers have been submitted electronically. Authors are advised to make preliminary email inquiries to editors about whether they are likely to be able to handle submissions in a particular electronic form.

***Algebra** to ROBERT GURALNICK, Department of Mathematics, University of Southern California, Los Angeles, CA 90089-1113; email: guralnic@math.usc.edu

Algebraic geometry to DAN ABRAMOVICH, Department of Mathematics, Boston University, 111 Cummington St., Boston, MA 02215; email: abramovic@bu.edu

***Algebraic number theory** to V. KUMAR MURTY, Department of Mathematics, University of Toronto, 100 St. George Street, Toronto, ON M5S 1A1, Canada; email: murty@math.toronto.edu

Combinatorics and Lie theory to SERGEY FOMIN, Department of Mathematics, University of Michigan, Ann Arbor, Michigan 48109-1109; email: fomin@umich.edu

Complex analysis and complex geometry to DUONG H. PHONG, Department of Mathematics, Columbia University, 2990 Broadway, New York, NY 10027-0029; email: phong@math.columbia.edu

***Differential geometry and global analysis** to LISA C. JEFFREY, Department of Mathematics, University of Toronto, 100 St. George St., Toronto, ON Canada M5S 3G3; email: jeffrey@math.toronto.edu

Dynamical systems and ergodic theory to ROBERT F. WILLIAMS, Department of Mathematics, University of Texas, Austin, Texas 78712-1082; email: bob@math.utexas.edu

***Functional analysis and operator algebras** to MARIUS DADARLAT, Department of Mathematics, Purdue University, 150 N. University St., West Lafayette, IN 47907-2067; email: mdd@math.purdue.edu

***Geometric analysis** to TOBIAS COLDING, Courant Institute, New York University, 251 Mercer St., New York, NY 10012; email: colding@cims.nyu.edu

***Geometric analysis** to MLADEN BESTVINA, Department of Mathematics, University of Utah, 155 South 1400 East, JWB 233, Salt Lake City, Utah 84112-0090; email: bestvina@math.utah.edu

Harmonic analysis to ALEXANDER NAGEL, Department of Mathematics, University of Wisconsin, 480 Lincoln Drive, Madison, WI 53706-1313; email: nagel@math.wisc.edu

Harmonic analysis, representation theory, and Lie theory to ROBERT J. STANTON, Department of Mathematics, The Ohio State University, 231 West 18th Avenue, Columbus, OH 43210-1174; email: stanton@math.ohio-state.edu

***Logic** to STEFFEN LEMPP, Department of Mathematics, University of Wisconsin, 480 Lincoln Drive, Madison, Wisconsin 53706-1388; email: lempp@math.wisc.edu

Number theory to HAROLD G. DIAMOND, Department of Mathematics, University of Illinois, 1409 W. Green St., Urbana, IL 61801-2917; email: diamond@math.uiuc.edu

***Ordinary differential equations, and applied mathematics** to PETER W. BATES, Department of Mathematics, Michigan State University, East Lansing, MI 48824-1027; email: peter@math.msu.edu

***Partial differential equations** to PATRICIA E. BAUMAN, Department of Mathematics, Purdue University, West Lafayette, IN 47907-1395; email: bauman@math.purdue.edu

***Probability and statistics** to KRZYSZTOF BURDZY, Department of Mathematics, University of Washington, Box 354350, Seattle, Washington 98195-4350; email: burdzy@math.washington.edu

***Real analysis and partial differential equations** to DANIEL TATARU, Department of Mathematics, University of California, Berkeley, Berkeley, CA 94720; email: tataru@ math.berkeley.edu

All other communications to the editors should be addressed to the Managing Editor, WILLIAM BECKNER, Department of Mathematics, University of Texas, Austin, TX 78712-1082; email: beckner@math.utexas.edu.

Titles in This Series

807 **Carlos A. Cabrelli, Christopher Heil, and Ursula M. Molter,** Self-similarity and multiwavelets in higher dimensions, 2004

806 **Spiros A. Argyros and Andreas Tolias,** Methods in the theory of hereditarily indecomposable Banach spaces, 2004

805 **Philip L. Bowers and Kenneth Stephenson,** Uniformizing dessins and Belyĭ maps via circle packing, 2004

804 **A. Yu. Ol'shanskii and M. V. Sapir,** The conjugacy problem and Higman embeddings, 2004

803 **Michael Field and Matthew Nicol,** Ergodic theory of equivariant diffeomorphisms: Markov partitions and stable ergodicity, 2004

802 **Martin W. Liebeck and Gary M. Seitz,** The maximal subgroups of positive dimension in exceptional algebraic groups, 2004

801 **Fabio Ancona and Andrea Marson,** Well-posedness for general 2×2 systems of conservation laws, 2004

800 **V. Poénaru and C. Tanasi,** Equivariant, almost-arborescent representations of open simply-connected 3-manifolds; A finiteness result, 2004

799 **Barry Mazur and Karl Rubin,** Kolyvagin systems, 2004

798 **Benoît Mselati,** Classification and probabilistic representation of the positive solutions of a semilinear elliptic equation, 2004

797 **Ola Bratteli, Palle E. T. Jorgensen, and Vasyl' Ostrovs'kyĭ,** Representation theory and numerical AF-invariants, 2004

796 **Marc A. Rieffel,** Gromov-Hausdorff distance for quantum metric spaces/Matrix algebras converge to the sphere for quantum Gromov-Hausdorff distance, 2004

795 **Adam Nyman,** Points on quantum projectivizations, 2004

794 **Kevin K. Ferland and L. Gaunce Lewis, Jr.,** The $RO(G)$-graded equivariant ordinary homology of G-cell complexes with even-dimensional cells for $G = \mathbb{Z}/p$, 2004

793 **Jindřich Zapletal,** Descriptive set theory and definable forcing, 2004

792 **Inmaculada Baldomá and Ernest Fontich,** Exponentially small splitting of invariant manifolds of parabolic points, 2004

791 **Eva A. Gallardo-Gutiérrez and Alfonso Montes-Rodríguez,** The role of the spectrum in the cyclic behavior of composition operators, 2004

790 **Thierry Lévy,** Yang-Mills measure on compact surfaces, 2003

789 **Helge Glöckner,** Positive definite functions on infinite-dimensional convex cones, 2003

788 **Robert Denk, Matthias Hieber, and Jan Prüss,** $\mathcal{R}$-boundedness, Fourier multipliers and problems of elliptic and parabolic type, 2003

787 **Michael Cwikel, Per G. Nilsson, and Gideon Schechtman,** Interpolation of weighted Banach lattices/A characterization of relatively decomposable Banach lattices, 2003

786 **Arnd Scheel,** Radially symmetric patterns of reaction-diffusion systems, 2003

785 **R. R. Bruner and J. P. C. Greenlees,** The connective K-theory of finite groups, 2003

784 **Desmond Sheiham,** Invariants of boundary link cobordism, 2003

783 **Ethan Akin, Mike Hurley, and Judy A. Kennedy,** Dynamics of topologically generic homeomorphisms, 2003

782 **Masaaki Furusawa and Joseph A. Shalika,** On central critical values of the degree four L-functions for GSp(4): The Fundamental Lemma, 2003

781 **Marcin Bownik,** Anisotropic Hardy spaces and wavelets, 2003

780 **S. Marmi and D. Sauzin,** Quasianalytic monogenic solutions of a cohomological equation, 2003

779 **Hansjörg Geiges,** h-principles and flexibility in geometry, 2003